Heaven's net casts wide.
Though its meshes are coarse, nothing slips through.
　　　　　—LAO-TSU

A Sierra Club Book

GALAXIES

Written and with photographs selected by TIMOTHY FERRIS

Illustrations by Sarah Landry

Stewart, Tabori & Chang, Publishers, New York

Frontispiece
The spiral galaxy designated NGC6744 is 300 million light-years distant from earth; a black-and-white photograph of this galaxy appears on page 109.

Copyright under the Berne Convention.
Published in 1982 by Stewart, Tabori & Chang, Publishers, New York.
All rights reserved. No part of the contents of this book may be reproduced by any means without the written permission of the publisher.

Distributed by Workman Publishing Company, Inc., 1 West 39th Street, New York, New York 10018.

Library of Congress Cataloging in Publication Data

Ferris, Timothy.
  Galaxies.

  Reprint. Originally published: San Francisco:
Sierra Club Books, c1980.
  Bibliography: p.
  Includes index.
  1. Galaxies.   I. Title.
QB857.F47 1982   523.1'12   81-21520
ISBN 0-941434-01-X   AACR2
ISBN 0-941434-02-8 (pbk.)

Printed and bound in Japan

Acknowledgments

The following are some of those who were kind enough to help out with *Galaxies*. Since none saw the entire book in its finished form, responsibility for errors or shortcomings remains solely that of the author.

The sun, the stars and seasons as they pass—some can gaze upon these with no strain of fear.
—HORACE

Information and Photographs

Halton Arp, Elly M. Berkhuijsen, Richard Berry, K. Alexander Brownlee, Lloyd Carter, S. Chandrasakhar, Mark R. Chartrand III, J. N. Clarke, James Cornell, A. G. de Bruyn, Terry Dickinson, Alan Dressler, Reginald Dufour, Vince Ford, Ken Franklin, Paul Gorenstein, J. Richard Gott III, Stephen T. Gottesman, John Graham, Edward J. Groth, B. W. Hadley, W. E. Harris, Eric B. Jensen, T. D. Kinman, Martha Liller, David Malin, Dennis Meredith, Simon Mitton, Richard Muller, Barry Newell, Rene Racine, Connie Rodriguez, D. H. Rogstad, Paul Routly, Ronald E. Royer, Vera Rubin, Allan Sandage, Jan Schafer, Malcolm Smith, Stephen Strom, Laird A. Thompson, Alar Toomre, Sindey van den Bergh, J. M. van der Hulst, Gerard de Vaucouleurs, Richard M. West, Fujiko Worrell, James D. Wray. Anglo-Australian Observatory, Astrophoto Laboratory, the Australian National University, Brooklyn College of the City University of New York, California Institute of Technology, Cambridge University, Carnegie Institution of Washington, the Cerro Tololo Inter-American Observatory, Dominion Astrophysical Observatory, European Southern Observatory, Griffith Observatory, Hale Observatories, Harvard College Observatory, Hayden Planetarium, Kitt Peak National Observatory, Lawrence Berkeley Laboratory of the University of California, Lick Observatory, Los Angeles Public Library, Max-Planck-Institut für Radioastronomie, McDonald Observatory, McMaster University, Mt. Stromlo and Siding Springs Observatories, New York Public Library, Princeton University, Rice University, Royal Observatory Edinburgh, Smithsonian Astrophysical Observatory, United States Naval Observatory, United States Naval Research Laboratory, Université de Montréal, the University of Chicago, the University of Florida, the University of Minnesota, the University of Texas at Austin, the University of Toronto, the Westerbork Radio Observatory.

Editorial Consultation
Timothy Crouse, Alan Dressler, J. Richard Gott III, Lynda Obst, Dennis Overbye, R. Bruce Partridge, Thomas M. Powers, Stephen Strom, Gerard de Vaucouleurs.

Advice and Encouragement
Jon Beckmann, Monica Brown, Jean B. Ferris, Wendy Goldwyn, Kathy Lowry, Lynda Obst, Bruce Porter, Thomas M. Powers, Delfina Rattazzi, Lisa Robinson, Allan Sandage, Alex Shoumatoff, Erica Spellman, Carolyn Zecca.

Illustrations
Sarah Landry

Research
Eileen Casey, Eustice Clarke, Juan de Jesus, Robert Ginsberg, Sandra Kitt, Judy Mitko, Terry Tammadge.

Dedication
To astronomers everywhere.

Table of Contents

Introduction

The Heavens ... are now seen to resemble a luxuriant garden, which contains the greatest variety of productions, in different flourishing beds; and one advantage we may at least reap from it is that we can, as it were, extend the range of our experience to an immense duration. For, to continue the simile I have borrowed from the vegetable kingdom, is it not almost the same thing, whether we live successively to witness the germination, blooming, foliage, fecundity, fading, withering, and conception of a plant, or whether a vast number of specimens, selected from every change through which the plant passes in the course of its existence be brought at once to our view?

—WILLIAM HERSCHEL

Where Are We?

REBECCA: *I never told you about that letter Jane Crofut got from her minister when she was sick. He wrote Jane a letter and on the envelope the address was like this: It said: Jane Crofut; The Crofut Farm; Grover's Corners; Sutton County; New Hampshire; the United States of America.*

GEORGE: *What's funny about that?*

REBECCA: *But listen, it's not finished: the United States of America; Continent of North America; Western Hemisphere; the Earth; the Solar System; the Universe; the Mind of God—that's what it said on the envelope.*

GEORGE: *What do you know!*

REBECCA: *And the postman brought it just the same.*

—*Thornton Wilder*

Children are forever reinventing the game of the Long Address. Like all enduring games it is earnest at the root, and its concerns pass into adulthood. We should like to be able to write a flawless version of the Long Address, to dispel its elements of fancy and replace them with facts, as in adulthood we replace the potential with the actual, anticipation with experience, and ourselves with our children.

We are much closer to that goal now than we have ever been before. Having learned our way around our own planet, we have extended the scope of human vision deep into the cosmos, to find that our world is but one of many worlds in one of many galaxies. Having come to view the broad universe on a scale approaching that of its construction, we can attempt to recite the Long Address in earnest. In its present state, it reads something like this:

The Earth. A small planet orbiting the sun, a yellow dwarf star.

The Solar System. Nine known planets plus a variety of smaller bodies, among them comets and asteroids, all orbiting the sun. The earth, third planet out from the sun, follows an orbit with a radius of some ninety-three million miles; light from the sun takes 8.3 minutes to traverse this distance, so we might say that the earth is 8.3 light-minutes from the sun. The outermost of the known planets, Pluto, reaches a maximum distance of 4.6 billion miles, or a little under seven light-hours, from the sun. Beyond Pluto lies the realm of the comets; when they are taken into account, the radius of the solar system may amount to as much as a few light-days.

The Sun's Neighborhood. Within seventeen light-years of the sun are sixty known stars, the nearest of them 4.3 light-years away. The solar neighborhood has been described by the astronomers Peter van de Kamp and Sarah Lee Lippincott as resembling "some sixty small spheres—tennis balls, golf balls, marbles and a large proportion of smaller objects—spread at random through a spherical volume the size of earth." Most of these neighboring stars are utterly unspectacular. Despite their proximity, fewer than a dozen—Sirius, Alpha Centauri, Procyon, Altair among them—are bright enough to be seen with the unaided eye in the skies of earth. Most are dim bulbs like Barnard's Star, Wolf 359, and BD +36° 2147. Here in nearby space we find affirmed a lesson of nature already known to naturalists who study beetles or bacteria, that some of the least conspicuous inhabitants of creation are likely to be its most numerous.

Vicinity of the Orion Arm. The solar system lies near a luminescent spiral arm of our galaxy. As the stars of the constellation Orion lie within the arm at approximately its nearest point to us, less than two thousand light-years away, we have designated it the Orion Arm. The arm is not an object but a zone where new stars have been created and have lit up the interstellar gas surrounding them, like luminescent plankton churned up in the wake of a ship.

The Milky Way Galaxy. The starry congress to which the sun belongs is a major spiral galaxy, a giant wheeling system roughly one hundred thousand light-years in diameter and home to better than two hundred billion stars. This is quite an abundance of stars; if we were to launch expeditionary forces at such a fantastic rate that an expedition reached a new star in our galaxy every hour of the day and night, and we kept up this rate of exploration year after year, we would have visited fewer than half the stars in the Milky Way Galaxy in ten million years, five times longer than the present tenure of our species. So large and abundantly populated by stars is our galaxy that I doubt whether anyone would feel disappointed had it proved to be the whole of the cosmos. But it is only one among many galaxies.

The Local Group of Galaxies. The Local Group, a small cluster of galaxies bound together gravitationally, is dominated by a pair of large spirals, the Milky Way and the Andromeda galaxies. Its radius is roughly three million light-years. A map of the group appears on page 85.

The Local Supercluster. The Local Group lies near the outskirts of the Local Supercluster, a vast aggregation of clusters of galaxies, its radius perhaps one hundred million light-years. The Local Supercluster is discussed in Section Five of this book. Some of the groups belonging to it are mapped on page 153.

The Universe. The population of the universe has been estimated at one hundred billion major galaxies, a figure whose tidiness suggests its inexactitude. The universe is said to be expanding and evolving. By expanding it is meant that the clusters of galaxies are rushing apart from one another at velocities proportional to their distances; by interpolating the rate of expansion backward in time we can infer that all the stuff of the universe was once crammed together at a very high temperature and density. In short, the cosmos did not always look the way it does today. It has changed throughout its history of eighteen billion years or so, and continues to change today. Once highly homogeneous and uniform, the universe has differentiated into a startling variety of forms, among them galaxies and their stars, planets and ourselves. It is this that we mean when we say that the universe is evolving. Since expansion and evolution are functions of time, we may wish to add a temporal dimension to our version of the Long Address. When we have done so, it reads:

The Earth
The Solar System
The Sun's Neighborhood
Vicinity of the Orion Arm
The Milky Way Galaxy
The Local Group of Galaxies
The Local Supercluster
The Universe, circa eighteen billion years
    after the beginning of its expansion.

It was at this point that a species arose upon our planet which found itself able to discover the galaxies.

The Discovery of Galaxies

Knowing as we do today that the universe is amenable to investigation, that our telescopes can examine millions of galaxies at distances of millions of light-years, we may be tempted to feel impatient at our predecessors for having subscribed to cosmological doctrines that we now understand—thanks in part to their efforts—to have been inaccurate. The problems that they faced were not minor, as a brief review of the discovery of glaxies will suggest.

Though they shine with the light of many billions of suns, most galaxies are so distant that they look faint. Only three galaxies are visible to the naked eye from the surface of the earth. These are the two Magellanic Clouds, which lie in southern skies, and the Andromeda Galaxy, whose tenuous glow was aptly described by a seventeenth-century observer as resembling "the light of a candle shining through horn." Dozens more galaxies may be seen with a small telescope. But, again owing to their great distances, they cannot readily be resolved into their billions of constituent stars. Only when giant telescopes were wed with cameras and sophisticated equipment such as the spectroscope could the stars of distant galaxies be discerned and analyzed. It was for this reason that galaxies were not discovered—that is, were not understood to be what they are—until the twentieth century. Prior to that, the question of the large-scale structure of the universe was approached primarily by way of studying our own galaxy, as a city dweller might learn something of the nature of cities by studying solely the city in which he lives.

The Milky Way, the disk of our galaxy as seen from our perspective within the disk, is composed of a multitude of stars sufficiently far away that our eyes cannot resolve them. Instead, we see only a general wash of light. Observers as early as Democritus speculated that the Milky Way was made of stars, but not until Galileo trained a telescope on the skies could the hypothesis be confirmed. "The galaxy is nothing other than a mass of luminous stars gathered together," Galileo wrote. He told the news to Milton, who accordingly wrote in *Paradise Lost* of "The Galaxy, that Milky Way/Which nightly as a circling zone thou seest/Powder'd with stars...."

Galileo's observations marked the beginning of the end of purely speculative cosmology. Prior to the invention of the telescope, when information about the stars was limited to what could be seen with the unaided eye, each theory of the structure of the universe had to be both concocted and judged almost entirely upon its merits as a creation of the imagination. Copernicus himself shared in this tradition, when on his deathbed he consented to publication of a cosmology that envisioned the earth orbiting the sun, for at the time the data on planetary motions fit the Copernican theory no better than they did the earth-centered cosmology of Ptolemy. Nor did the tradition of speculative cosmology die with the advent of Galileo. It continued at least until the eighteenth century, when it enjoyed a baroque efflorescence in the speculations of the young Immanuel Kant.

During the century between Galileo's death and Kant's boyhood several observers equipped with telescopes had noted the presence among the stars of ill-defined, luminous patches they called "nebulae." Today we know that several different sorts of objects were lumped together under the term. Most belonged to our galaxy; among them were clouds of gas illuminated by stars within them, shells of gas ejected by aging stars, a few indistinct clusters of stars. But some —the spiral nebulae— were galaxies in their own right, independent of the Milky Way.

It was Kant who guessed correctly that the spiral nebulae were galaxies. In 1755, when Kant was thirty years old, he wrote, "If a system of Fixed Stars which are related in their positions to a common plane, as we have delineated the Milky Way to be, be so far removed from us that the individual stars of which it consists are no longer sensibly distinguishable even by the telescope; if such a World of Fixed Stars is beheld at such an immense distance from the eye of the spectator situated outside of it, then this World will have the appearance of a small patch of space whose figure will be circular if its plane is presented directly to the eye, and elliptical if it is seen from the side or obliquely. The feebleness of its light, its shape, and the apparent size of its diameter will clearly distinguish such a phenomenon, when it presents itself, from all the stars that are seen as single." Word for word, it would be difficult to improve upon this description of the appearance of spiral galaxies.

Kant's cosmological speculations, published anonymously by a publisher who promptly went out of business, passed unnoticed at the time. Even had his theory won attention, there would have been at the time no way of putting it to an observational test. And that is the difference between the speculative cosmology of previous epochs and observational cosmology as practiced today. The vastly improved telescopes and other instruments now available to scientists permit them to test cosmological speculations by direct observation. The technological advances involved may be summarized in three categories—the use of the spectroscope to study the physics of stars, the development of precision telescopes able to measure the distances of nearby stars, and the building of giant telescopes with light-gathering power equal to the task of investigating remote galaxies.

The spectroscope enables researchers to examine the anatomy of light. The spectrum it produces might be compared to the musical score employed by a conductor to examine the parts played by individual musicians in an orchestra. The atoms of each element generate energy within a characteristic range of frequencies, as do the instruments in an orchestra, and within that range each atom plays a variety of melodies and harmonies from the study of which can be learned an enormous amount about the state of the atoms and of their environment. And if a star is moving toward or away from us, the frequency of the notes played by each of its atoms will be altered—shifted toward a higher frequency if the star is moving toward us, toward a lower frequency if it is moving away from us, in much the same way as a car horn sounds a different pitch if the car is approaching or receding. By analyzing this effect, known as the Doppler shift, astronomers can determine how rapidly a star is rotating, how fast it is moving in space, the extent of the churning motions within an interstellar cloud surrounding it, the velocity of stars in their orbits in another galaxy, and the velocity of galaxies as they participate in the general expansion of the universe. The uses of spectroscopy are almost limitless.

Just as the spectroscope helped to answer the ancient question of what the stars are made of, the equally fundamental question of the distances of stars was approached through the rise of precision astronomy in the nineteenth and early twentieth centuries, by way of the method of parallax. The fundamental principle of parallax is that the distance to a nearby star can be measured by changing our perspective on it. Hold your index finger at arm's length, close your left eye and sight the finger against a distant background: then close your right eye, open the left, and notice the apparent shift in the position of your finger; that is parallax. In astrometry—the precise measurement of the apparent positions of stars—the shift in perspective may be provided by taking two photographs of a relatively nearby star at an interval of six months, time enough to allow the earth in its orbit around the sun to shift its position by some one hundred eighty-six million miles, altering our perspective on the star the way switching from the right to the left eye altered our perspective on our finger. The diameter of the earth's orbit is not a very large distance by the standards of the stars, and astronomers working at the limits of this parallax method are obliged to measure shifts in perspective as minute as that which would be obtained by squinting alternately with one's right and left eyes at a finger six hundred miles away. But given an almost obsessive dedication to precision, reasonably accurate distances may be obtained for stars as far away as several hundred light-years.

Distances greater than a few hundred light-years are usually determined by estimating the instrinsic brightness of a star —known as its absolute magnitude—and comparing that value with its apparent brightness in the sky, known as apparent magnitude. The apparent magnitude of a star or other astronomical object decreases with the square of its distance, so determining its distance is simple once one knows how bright the star really is.

Astronomers have developed a variety of ingenious ways of estimating the absolute magnitudes of stars, most of them involving extrapolation from the studies of nearby stars whose distances could be determined by the parallax method. Nature has obliged them by creating variable stars of a sort whose period of brightness variation is directly related to their absolute magnitude. Once one knows the absolute magnitude of just a few such variables, one can use their cousins as distance indicators deep into the reaches of space. Particularly helpful in this regard have been the Cepheid variable stars, so bright that they can be identified in galaxies well beyond our own. Cepheids pulsate in brightness at intervals of as little as one day or as much as seventy days or more; the period of any given star betrays its absolute magnitude. Its distance can then be derived by comparing its absolute magnitude with its apparent magnitude. Distances to a number of neighboring galaxies, among them the giant spirals M81 and the Andromeda Galaxy, have been estimated by examining their Cepheid variable stars.

But Cepheid variables cannot be detected with existing telescopes at distances much greater than ten million light-years. To go farther, astronomers measure the size and brightness of glowing gas clouds and clusters of brilliant supergiant stars in galaxies, then estimate the distances of those galaxies by proceeding on the assumption that these denizens of other galaxies resemble their like here in the Milky Way. Explosions of stars as novae and supernovae also can be observed in distant galaxies and employed as a check against estimates made in other ways.

Beyond a few tens of millions of light-years the efficacy of these methods fades, and astronomers fall back upon the brightness and diameter of whole galaxies as indices of their distance. Here the assumption is that the dominant spiral in each cluster of galaxies will prove on the average to be comparable to that of dominant spirals like the Andromeda Galaxy in our Local Group.

Finally, the clusters of galaxies are rushing apart from one another as the universe expands; their velocities can be ascertained by measuring Doppler shifts in the spectra of their light. The farther away any given cluster of galaxies, the more rapidly it is receding in the expansion of the universe. (This is discussed more fully in Chapter Five.) So the distances of galaxies hundreds of millions or even billions of light-years away can be inferred from their velocity of recession.

That we can observe distant galaxies at all, much less determine their distances, analyze the compositions of their stars and interstellar clouds, measure their rates of rotation and from that estimate their mass, and otherwise chart them with a precision that few scientists a century ago would have thought possible—all this is due primarily to the advent of large telescopes. Unlike their colleagues in other physical sciences, astronomers cannot probe or dissect or experiment with the objects of their attention. Starlight washes down over the earth, a gentle rain indeed. All that the astronomer can do is to gather a little of it, bring it to a focus, and subject it to examination—by spectroscope, photometer, photographic plate or electronic detector. The more cosmic energy that can be gathered, the better, whether in the form of light, natural radio emanations, infrared and ultraviolet light, or the high-frequency energies of cosmic rays, X-rays and gamma rays. And more cosmic energy has been gathered and analyzed in our century than in all human history preceding it. This development more than any other has elevated our knowledge and awareness of the wider scheme of things.

At the vanguard of this scientific revolution was the construction by the American astronomer George Ellery Hale of large telescopes at the Mount Wilson and Palomar Observatories in California. It was at Mount Wilson that Harlow Shapley ascertained that the sun is located not near the center of the Milky Way Galaxy, as many had thought, but toward the outskirts of its disk. At Mount Wilson in 1924 Edwin Hubble determined that there are galaxies beyond ours, and that their stars are similar to stars found in our galaxy. Thus, questions concerning the existence of galaxies and our place within the Milky Way were transferred from the realm of speculation to the realm of verifiable fact.

From the work of Shapley, Hubble and their colleagues emerged two profound revelations about the nature of nature at large. The first is that the universe is far larger and more various than almost anyone had been prepared to believe. The other is that the breathtaking scope and diversity of the universe has arisen within the constraints of principles of nature—physical laws, as they are called—the same as those that pertain here at home. Nature everywhere plays by the same rules, and by learning those rules we can learn from nature, everywhere.

This second principle makes possible the science of astrophysics—the application of physical principles learned in laboratories here on earth to phenomena beyond the earth. We can estimate the mass of galaxies from their rotational velocities and from their interactions with one another because they obey the same principles—those elucidated by Newton and Einstein—as do falling apples on earth and the orbits of planets in the solar system. We can determine the composition of distant stars because they are made of the same sort of atoms as those we find here on earth or in the sun.

A third fundamental discovery of twentieth-century astronomy has been that of the expansion and what might be called the evolution of the universe. In 1929 Hubble, working in part from data supplied by a fellow astronomer, Vesto Slipher, discovered that remote galaxies are rushing apart from one another. Subsequent refinements in measurements of the rate of expansion have led to a modern estimate of roughly eighteen to twenty billion years since expansion began. Indirect confirmation of this time scale has come from the dating by astrophysicists of the oldest stars at some fifteen billion years of age, and from the discovery by radio astronomers of the cosmic background radiation, residual energy left over from the fiery moment when expansion began, its characteristics in accord with those of an expanding universe eighteen to twenty billion years of age.

Evolution, a word admittedly fraught with ambiguity, comes into play if we consider how the variety and diversity of the cosmos has increased with the passage of time. At the instant of the "big bang," any one scoopful of the stuff of the cosmos would have closely resembled any other—each would be essentially a batch of pure energy. Soon after expansion began much of the energy cooled to form the primordial elements hydrogen and helium, so that a token scoop would contain a greater variety of particles—hydrogen, helium and photons of energy—but each scoop would still generally resemble any other. Today the diversity of the cosmos is so great that it is probably fair to say we have not begun to imagine it, much less to observe it. There are billions of galaxies, each with myriad varieties of stars and untold numbers of planets whose diversity of detail may perhaps be hinted at by the variety of life on earth and by our human thoughts about the cosmos. A scoopful of the cosmos drawn at random today might hold empty space, the alcohol molecules of an interstellar cloud, a dry ice snowball like those found on Mars, a rabbit's foot or words in a book. We see the universe today as a dynamic system within which human evolution plays a small but perhaps not discordant part. Herschel's vision of starry gardens where we find all sorts of plants at various stages in their lives has never seemed more nearly true.

About The Photographs

Writers who report on science solely in terms of its results are like hunters who shoot leopards solely for their skins. I am guilty of just such reporting in this book. I have presented the results of science, with little mention of the astronomers and astrophysicists whose work produced these results and made the book possible. I hope that they will not mistake this for ingratitude. My aim has been to encourage us to look at the galaxies directly, to appreciate that they are not solely specimens arrayed for scientific study, but that they are part of—most of—the natural world, as real and as worthy of our attention as are we who behold them. To do this I have been obliged to treat the scientific process as but a window onto the galaxies. I have tried to make that window so clear that we might occasionally forget that it is there.

The photographs in this book were taken, with a few exceptions, by astronomers employing some of the largest telescopes at various observatories around the world. The names of the observatories appear on page 5. There are many methods of depicting galaxies other than by photographs taken at optical wavelengths, but the human mind is strongly oriented around the visual sense, and so I have limited non-optical images to a few radio plots of galaxies and to the X-ray quasar image in Chapter Five.

The photographs are time exposures, obtained by aiming a telescope at a galaxy and exposing a photographic plate for as long as several hours while starlight seeps into the photographic emulsion. During this time a driving mechanism compensates for the earth's rotation and keeps the telescope trained on the galaxy, while the astronomer, or in some cases an automatic guiding system, makes minute corrections to compensate for refraction of light by the atmosphere or for inaccuracies in the driving mechanism.

The resulting photographs inevitably represent various compromises. Some compromise is involved in deciding upon exposure time. The interior sections of spiral galaxies contain a far higher density of stars and so are much brighter than the disks. A photograph that shows the spiral arms in detail must therefore overexpose the central region, while a photograph made to study the central regions will record little of the spiral arms. One can compensate for this by making two exposures, one for the arms and one for the central region, and combining them, but the result will produce an inaccurate impression of the galaxy's brightness profile. Another compromise involves the color sensitivity of the film chosen: a film preferentially sensitive to the red end of the spectrum will better record the ruddy bright nebulae that lie along the Milky Way, while a blue-sensitive film will suppress the clouds but bring out the young stars that lie within them. Astronomical photography, like all photography, contains an element of the impressionistic. At a few points I have endeavored to compensate for these limitations by presenting several photographs of the same galaxy taken in various wavelengths of light, as in the case of M82 (page 136), or by presenting detailed photographs of specific regions of a galaxy, as with the Andromeda spiral (pages 77-79).

The color photographs were in most cases produced by exposing three black and white plates, each limited by filters to one band of the spectrum, then combining them to produce a finished three-color print. The colors of galaxies cannot be seen directly by the human eye, even through the largest existing telescopes, for their light is too faint to stimulate the color receptors of the retina. And, since some elements of human judgment inevitably are involved in the production and reproduction of color photographs, slight differences in color balance may result when two observers create color images of the same galaxy. But the colors themselves are real, and the photographs represent the best efforts of astronomers to reproduce them accurately.

Galaxies and Human Thought

The study of galaxies by human beings has scarcely begun. Someone a century from now, reading what we thought about galaxies, would no doubt find much of it distorted, stunted or simply wrong. A map like the one that appears on page 42, which hazards to depict the sun's surroundings in our quarter of the Milky Way Galaxy, may by then seem as quaint as sixteenth-century maps of the New World, with their many errors and broad swaths of *terra incognita*. If our descendants smile at our ignorance, perhaps they will understand that we would have welcomed its banishment no less enthusiastically for its having been our own.

Looking back on our century they may be tolerant of us, keeping in mind that the discovery of galaxies and of the enormity and diversity of the cosmos came as something of a shock. We have lost the cosmos of our forebears, where the sky was wrapped around the world like a blanket and we walked upon a land that had been created for ourselves alone. It has not been easy for us to appreciate that there are many skies wrapped around many worlds and that the galaxies that harbor these worlds are real, that their fiery stars and planets and the ghostly clouds that waft among them are as much a part of the natural world as is any sun-basked meadow here on earth. It is not surprising that many of us feel nostalgia for older cosmologies, and that from this have sprung dissatisfied reactions to the new. One such reaction concerns location: Some feel remorse that we are removed from what we had imagined was a throne at the center of the universe. Another concerns diminution: How can we tiny creatures retain our self-esteem after having been exposed to the grandeur of the universe at large? A third concerns change: If the earth, stars and galaxies were born, are changing and will one day die, upon what rock may we build a church of any permanence?

We might be tempted to dismiss these objections by arguing that it doesn't matter how we feel about the galaxies, that their existence is a fact, and that we should accept that fact inasmuch as the secret of learning is to love the truth. But the existence of these feelings is fact, too, and most of us share in them to some degree. Stolid indeed is the student of galaxies who has felt no sense whatever of dislocation, disenfranchisement or vertigo at the sight of them. So those of us who feel that the benefits of the discovery of galaxies outweigh any trauma it may have produced might do well to explain why we feel that way.

As to location: It is true that we do not occupy the center of the universe. It seems that there *is* no center of the universe —except perhaps in the highly technical sense of a null point that might be discernible in the flux of the cosmic background radiation—and if such a center were located, one can scarcely imagine any reason why we would want to live there. Nor is it clear that any distinction ought to have been bestowed upon our imaginary occupancy of the center of things in the first place; the center of the world of Christian cosmology, for example, was occupied not by God or the angels but by Satan.

The universe, we now realize, is highly equitable in terms of location. The view is splendid from almost everywhere. If the sun were located in another galaxy we could just as easily observe the universe at large, and take photographs of galaxies such as the ones that appear in this book, upon one page of which would appear the Milky Way Galaxy. Here we are in the solar system, squired on the arm of a magnificent spiral galaxy; surely this is not grounds for complaint.

As to dimension: It is true that we are tiny relative to the cosmos. *Everything* is tiny relative to the cosmos—even a galaxy is but one among billions—and to fret about this is to confuse size with stature. We are well advised to bow to no tyranny of mere size, to heed the lesson of Lao-tsu, Aristotle, Leonardo and Darwin, who teach that the truth is less often attained by gaping at the grand than by scrutinizing the small. The human body is a galaxy to a microbe, yet without microbes the body would not live an hour. If we feel awe, let us address it not to dimension but to being—and we share being with the galaxies themselves.

As to change: The stars, once thought of as symbols of constancy, are seething fireballs hurtling along in a changing galaxy in a changing universe, and this realization clearly has its unsettling side. But so did the illusion of a changeless cosmos that preceded it. When the heavens were thought to be eternal and unalterable, it was tempting to regard the turmoil of human affairs as fundamentally different in character from the workings of nature. To live on earth was to be condemned to mutability and corruption, while the stars revelled in immutability and incorruptibility. The fundamental error of imagining that human life was fundamentally different from the rest of nature gave rise to many an odd doctrine. Aristotle argued that only the heavens were pure, Plato that the real world is static and archetypical and the flux of perceived existence but an illusion. But, then as now, others were willing to conceive of themselves as immersed in a changing cosmos. If Plato in his Olympian home is today disturbed by the cosmos in flux we believe we have discovered, his predecessor Heraclitus, who saw change in everything, warms his hands on the fires of the galaxies.

The reassuring aspect of the portrait of the universe we now see drawn across the sky lies in its reconciliation of humanity with the material world. That we are part of our galaxy is literally true. The atoms of which we are formed were gathered together in the toilings of a galaxy, their fantastical assembly into living creatures was nourished by the warmth of a star in a galaxy, we look at the galaxies with a galaxy's eye. To understand this is to give voice to the silent stars. Stand under the stars and say what you like to them. Praise or blame them, question them, pray to them, wish upon them. The universe will not answer. But it will have spoken.

T.F.

1/The Milky Way:

A Spiral Galaxy Viewed from Within

We too once lived in this house of stars....
—MURMURS OF EARTH

A Journey to the Center of the Milky Way

Our sun and its planets lie in the environs of the Milky Way Galaxy. To go to the center of the galaxy would require navigating a distance of some thirty thousand light-years. Such a journey lies far beyond the technological capacity of our species in this century; interstellar distances are vast, the energy required to traverse them enormous. Some say we shall never be able to do it. Others say we might. No one expects that we shall do it soon.

Yet we can make the trip today, aboard the ship of the imagination. This may seem like mere daydreaming, but dreams have preceded earlier journeys, as when our forebears contemplated oceanic horizons in the days before we mastered the seas. And the sights to be seen during such a journey need not be pure imagination; we have learned enough about the galaxies to predict in general terms what we might hope to see if we traveled through them.

If we need further encouragement, let us consider the remarkable accommodation of science to fantasy represented by the time-dilation effect in Einstein's special theory of relativity. The theory, verified in many an experiment, tells us that the passage of time slows down dramatically aboard a spaceship that is accelerated to velocities approaching that of light. (The speed of light itself can never be attained, it adds.) We can spend energy to buy time.

Owing to this effect a starship able to maintain an acceleration equal to the force of gravity here on earth could reach the center of our galaxy, thirty thousand light-years distant, in under twenty-five years of on-board time. Neighboring galaxies could be achieved in less than thirty years, and gulfs separating clusters of galaxies crossed in perhaps a decade longer. So let us imagine ourselves aboard such a ship, and see where it might take us.

The ship's appointments may be left to the preference of each passenger. We might envision a giant vessel complete with baseball teams, string quartets, a hardwood copse and trout pond, and a crew of thousands selected from backgrounds sufficiently varied to insure that things will never quite run smoothly. Or a more modest cruise starship, with a tiny nightclub, an indefatigable recreation director, many outside cabins with portholes. Or a military starship, all drums, boots and salutes. Each to his own. There is ample room in the imagination for imaginary starships, as there is in the cosmos for real ones.

The day of our departure is sad, its farewells permanent. We travelers will be able to take advantage of the time-dilation effect. Friends and families who stay at home will not. They will have been dead for tens of thousands of years by the time we reach the center of our galaxy. Together we sing the anthem of interstellar explorers everywhere, a song of final farewell. Then we depart.

The early years of the voyage pass uneventfully while our ship accumulates velocity. Years pass before we can celebrate having attained the distance of the nearest extrasolar star, Alpha Centauri, a little over four light-years out. The sun is now but a dot of light in the constellation Taurus. Soon it will become difficult to identify the dim little sun in the sky.

In the years that follow, the stars crawl across the sky, slowly distorting the constellations we have known on earth until most are unrecognizable. Our course takes us along the plane of our galaxy directly toward its center. Our view is composed of stars and of the clouds of dust and gas that lie in the spaces between the stars. The interstellar clouds are mostly dark, but when we encounter one of the spiral arms of our galaxy, we find them lined with bright nebulae—regions where newly formed stars have lit up the surrounding clouds— and the sight of these glowing shoals cheers us as we speed on.

Many a flower of these starry pastures could hold our attention—the high density dwarf stars, neutron stars and black holes, the endlessly varied matchups of multiple stars, the variable and flare stars, and the billions of ordinary stars like our sun, not to mention their planets. But we must hurry on.

After decades of travel, the interstellar clouds at last fall away. Ahead lies the central region of the galaxy, an elliptical cosmos of stars glowing through the relatively unsullied spaces with fantastic clarity. The color of this great egg is that of a bloodied yolk, the red and orange light of old stars that have been burning steadily for billions of years. Behind us the inner portions of the dusty disk hang like the walls of a canyon; thousands of light-years down one wall we can discern the elbow joint where one spiral arm emerges from the central regions and begins a winding path that will eventually take it out past our sun.

We plunge into the central regions. Country dwellers on our first visit downtown, we are surprised at the congestion of the stars, their hustling pace. Their light is as warm in hue as torch light. They pursue jitterbug orbits that seem hurried by the standards of the solar region, and the clearances among them are narrow. Yet all conduct their affairs without colliding.

Our destination is the nucleus of the galaxy. We can see its brilliant lamp ahead. What will we find there? An enormous star cluster, sitting like a clutch of diamonds at the center of the galactic diadem? The ominous warren of a black hole, a creature out of the Inferno rather than the Paradiso?

It is just at this point of the voyage of our imagination that we must turn away. We know that our galaxy has a nucleus, but do not yet know enough about it to be able to describe it. The captain orders our course altered, and our ship describes a sweeping arc that takes it up and out of the plane of the galaxy. Ahead lies intergalactic space.

1/THE MILKY WAY

The Milky Way is an old enchanter of the human sight. The abundant historical references to it tend to be fond, or grand, or both. To the ancient Chinese and Arabs, this softly glowing band of light resembled a river in the sky. Ovid and the Pythagoreans compared it to a bridge. It was a roadway to the Anglo-Saxons, an avenue leading to Valhalla for the Norse. The ancient Greeks compared it to milk, and it is from the Greek for "milk," *gala*, that our word "galaxy" is derived.

In a sense, the Milky Way has indeed turned out to be a kind of bridge or pathway leading our minds from earth into the heavens. It represents our view, from within, of our own galaxy, one among many galaxies in the universe. The discovery of that fact set us upon a pathway of discovery that promises to lead us out of our cosmological childhood.

The soft light of the Milky Way comes from billions of stars. Its definition into something resembling a path or river results from the fact that our galaxy, like any normal spiral galaxy, is flattened in form, the majority of its stars concentrated along a disk that is as thin in relation to its diameter as a heavy old coin. The light of the Milky Way is more intense in one direction, toward the constellation Sagittarius, in the southern skies of earth, for in that direction lies the center of our galaxy. Dark rifts that meander through the Milky Way like sinew through muscle are, we now know, dark clouds of dust and gas that block the light from the stars that lie beyond them. Bright clouds like the Orion Nebula we now understand to be part of a pyrotechnical display that illuminates the spiral arms of our galaxy; by mapping these nebulae we have been able to chart the parts of the arms of our galaxy that lie in our celestial neighborhood.

Our predecessors were right to have responded to the almost subconsciously embracing quality of the Milky Way and to have spoken of it in terms of things that are important to us here at home. The Milky Way Galaxy does embrace us, and is our home.

The Sun

We who inhabit earth enjoy an intimate relationship with one star, the sun. The first fact of this relationship is proximity, by virtue of which the sun, though only an average star, outshines all the other stars of our skies combined. The sun pursues its orbit as part of the wheeling of the Milky Way Galaxy, and we accompany it.

Planets are not terribly imposing members of the solar system, and the earth is one of the less imposing of the planets. Over ninety-nine percent of the mass of the solar system is the sun. Most of the remaining mass is bound up in the planet Jupiter. The other major planets—Saturn, Uranus and Neptune—account for nearly all of the rest of the mass. Finally come the lesser planets Mercury, Venus, Mars, Earth and Pluto, plus a few dozen satellites and a variety of asteroids ranging in size from rocks larger than the island of Manhattan down to particles smaller than grains of sand. The earth amounts to less than one-hundredth of one percent of the mass of the solar system.

But however incidental the planets may seem compared to the grandeur of stars, each planet retains grandeur in the human eye. Each is a world of its own, as we are coming to realize better by way of reconnaissance of the solar system by space probes. We have glimpsed the wind-blasted flatlands of Venus, the snow-capped ochre mountains of Mars, Jupiter with its fiesta-costumed moons a miniature solar system in itself, and Saturn, another miniature solar system, with its rings the colors of plum and sand, yin to Jupiter's yang. The beauty and bounty of the solar system, its great reservoirs of diversity, offer us a foretaste of what we may find when we are able to investigate elsewhere in our galaxy in detail. It is as yet uncertain what percentage of the stars have planetary systems. But even if only one star in ten thousand has planets, that would add up to ten million planetary systems in a galaxy the size of ours. When contemplating galaxies, we do well to keep in mind what a wealth of worlds each may harbor.

Our proximity to the sun makes us beneficiaries of its energy. The surface of the earth intercepts only a tiny fraction of the sun's energy, but this has been sufficient to make possible, among other things, the origin and evolution of life here. It is with some interest, then, that we inquire about the way in which this energy is produced.

The nuclear physics of the interior of a star involves the fusion of the nuclei of atoms and the release, in the process, of a fraction of the energy that binds each nucleus together. It is a sort of game of musical chairs in which, following certain rearrangements of subatomic particles, bits of energy

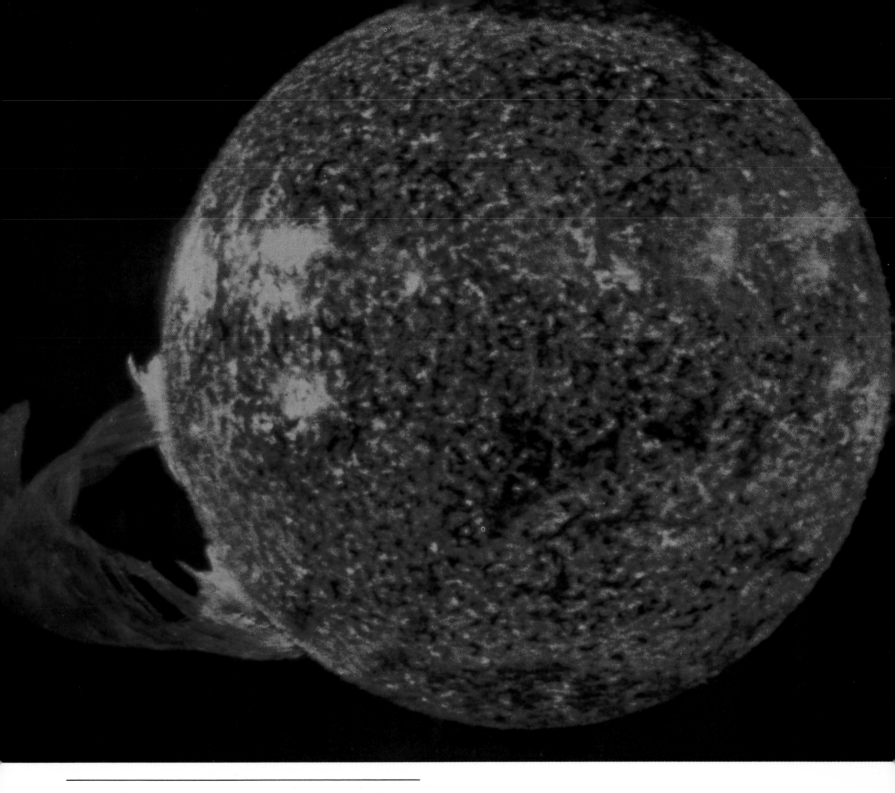

1 The sun during a major solar eruption, photographed in ultraviolet light from the Skylab orbiting space station in 1973.

Figure 1

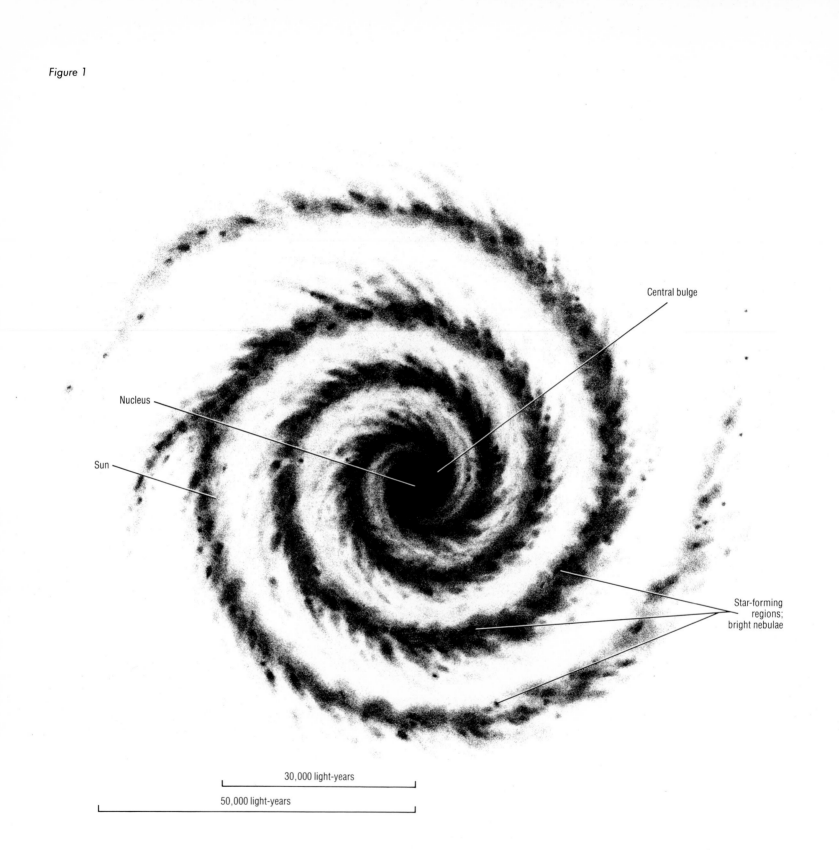

Central bulge

Nucleus

Sun

Star-forming
regions;
bright nebulae

30,000 light-years

50,000 light-years

Figure 2

Figure 1, 2. Milky Way Plan and Side Views
The anatomy of a normal spiral galaxy is presented in these views of
the Milky Way. The galaxy is centered on a compact nucleus surrounded
by a roughly spherical realm of stars usually referred to as the central
bulge. A spherical halo made up of scattered older stars embraces the
entire galaxy; here are to be found many globular clusters. Most of the
interstellar gas and dust of the galaxy, as well as most of its stars,
occupy the flattened disk. The spiral arms represent portions of the disk
made evident by the hosts of brightly shining new stars recently formed
there. The spiral arms are depicted schematically, as our galaxy has not
yet been well mapped beyond the solar neighborhood.

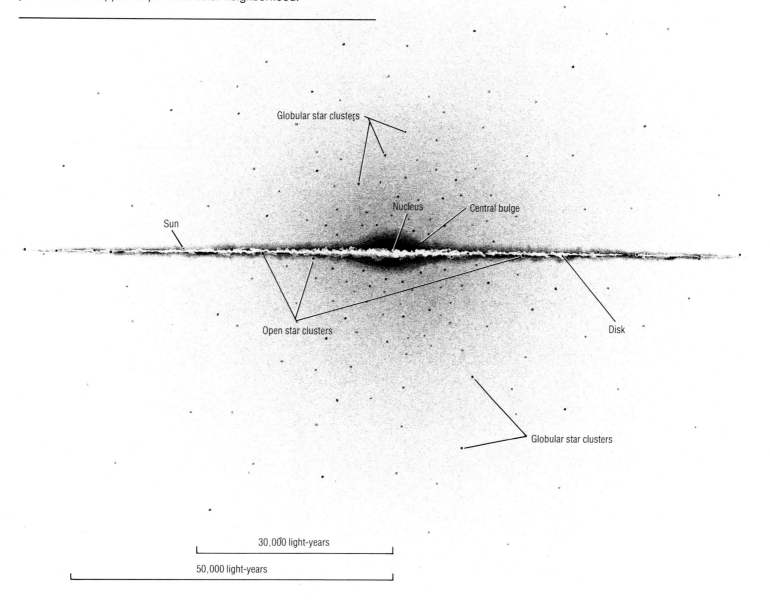

find themselves with nowhere to sit and so depart. In this fashion, the sun in its core converts nearly five million tons of matter into energy each second. The energy that has lost the game of musical chairs makes its way slowly—this takes millions of years—to the surface, where it is radiated into space. It is a matter of perspective whether we care to think that stars employ nuclear fusion to release energy, or that they fuse atomic nuclei and produce energy as a byproduct.

Many stars vary in brightness over periods of days or months, some wildly so. Fortunately for we who depend upon its even temper, the sun is at most only mildly variable. On the basis of historical and climatological records, the sun seems to have altered its energy output somewhat in the long term. In the short term, its most spectacular signs of distemper take the form of periodic eruptions on the surface that hurl solar material out into space.

The photograph (page 21) shows one unusually large eruption from the surface of the sun. Solar material, primarily hydrogen gas—in the cosmos one is forever encountering hydrogen gas—is being ejected like a bubble bursting from the surface of a whitewater brook. The granular appearance of the solar disk results from the action of convection currents, hot towers of upwelling material alternating with descending cells of cooler material.

The photograph was made in ultraviolet light. The energy emitted by the sun comes in a variety of wavelengths, including ultraviolet and infrared light; it even manages some output in the very long wavelength radiation we call radio. But it radiates most strongly along the portion of the electromagnetic spectrum we call visible light. This is no coincidence; we evolved on a planet bathed in sunlight, and our eyes evolved as to make the most efficient use of its energy.

Stars and Interstellar Space

Nature plays conjurer's tricks, producing endless diversity from the most ordinary of ingredients, pulling from its top hat not only rabbits and bouquets and conjurers themselves, but stars. Stars are made from the simplest of ingredients, hydrogen gas (and hydrogen is the simplest of the elements) mixed with some helium (the second simplest element), and traces of more complex atoms. A congealing gas cloud becomes a star once it has compacted itself to a sufficiently high density that the heat generated in the crush of its core rises to a point high enough to fuse atomic nuclei and release energy. Gravitational force, the universal attraction of matter

for matter, tends to collapse the star. Energy bubbling up from its core tends to push it apart. The balance between these two forces maintains a star's composure. A star goes through various changes as it ages, yet these changes are dictated almost entirely by how much mass it started out with; for instance, very massive stars burn more rapidly and age more quickly than stars of average mass.

Stars are remarkable in their variety. There are stars smaller than earth and stars larger than earth's orbit, stars younger than human civilization and stars nearly as old as the universe, stars harder than diamond and gasbag stars so diffuse that much of them is thinner than air, hot blue stars and dim stars that glow the ruby hue of a cooling coal, variable stars that pulse like jellyfish, flare stars that brighten as suddenly as a campfire doused with kerosene, single stars like the sun, and double stars, triple and quadruple stars.

The stellar population of our galaxy is estimated at something over two hundred billion. The photograph (right) shows a few of them. Their apparent crowding is an illusion created by the fact that we are seeing thousands of light-years into the depths of space; stars that appear to be stacked almost next to one another actually are separated by many light-years. Stars generally have plenty of room; a few dozen baseballs scattered across all North America would crowd one another in comparison with the leeway enjoyed by an average star in an average galaxy.

The stars we see in our galaxy also vary widely in age. Some are extremely old, some very young; most, like our sun, lie between those extremes. It follows that we ought to find evidence of stellar birth and death going on around us, just as in the human community we find that some persons are younger, some older, and that births and deaths are constantly taking place. And considerable evidence has accumulated that such is the case. In our galaxy may be found new stars being born and old stars in their death throes. The sites of some of these events appear on the following pages.

The spaces between the stars of our galaxy are littered with clouds of dust and gas. Stars form from such clouds, which are characteristically very thin, typically thinner than a

---

2 The Milky Way in its tangle of stars and interstellar clouds offers us a firsthand look at how a spiral galaxy—in this case, our own—appears from a vantage within and toward one side of the galactic disk.

laboratory vacuum, but so vast that they have enough mass in the aggregate to make billions of suns.

At any given time most of the atoms in an average interstellar cloud drift along on their own. Some couple with other atoms to form molecules, and molecules in turn wander the wastelands of space. For stars to form from such a cloud, enough of these wandering atoms must be brought together so that gravity, a very weak force, is able to tether them and arrest their independent meanderings. Once this has happened, the bundle of atoms that results is able to capture other atoms that it encounters, binding them to the group, slowly increasing the group's mass and with it its gravitational attraction. Star seeds like these are growing today in many places in our galaxy, each scarcely noticeable yet awesome in its potential, like an embryonic cell in a womb.

Interstellar clouds are generally dark and inconspicuous except when illuminated by stars that have just formed there, or where silhouetted against a background of stars. The two clouds pictured here are seen in silhouette.

Many different sorts of terminology can be applied to interstellar clouds. Most convenient is to lump them under the term "nebula," from the Latin for "mist," or "cloud." Illuminated interstellar clouds are called bright nebulae. Unilluminated clouds are called dark nebulae.

4

3 The dark interstellar cloud known as the Coalsack, nearly six hundred light-years away and seventy light-years in diameter, is nestled against the foot of the Southern Cross, which lies on its side in this photograph (left). In reality, only Beta Crux and Delta Crux, the stars at the ends of the short arm of the cross, occupy the same celestial neighborhood as does the Coalsack. Alpha Crux, the bright star at the foot of the cross, and Gamma Crux, the star at its head, are both in the foreground, at respective distances of about three hundred seventy and two hundred twenty light-years.

4 The Cone Nebula, twenty-six hundred light-years distant, is part of an extended interstellar cloud. The bright stars in the background apparently formed out of the cloud recently in cosmic history; their light now silhouettes the foreground part of the cloud.

Stars Being Born

THE ORION NEBULA

The first stars to form within a condensing cloud bestow the gift of light upon their progenitor. Previously dark, the cloud now bursts forth with a bouquet of color that makes bright nebulae among the most arresting sights in the sky. Some of the illumination consists of light from the young stars reflected by dust grains in the surrounding cloud. But most of it is produced when gas in the cloud, far more abundant than dust, is ionized—electrically charged—by starlight, and glows by reradiating the received energy, as does the gas in a neon light.

Bright nebulae like these are found lining the arms of spiral galaxies, where density waves set up by the rotation of the galaxy have recently passed through, promoting the condensation of interstellar clouds into stars. The sun currently finds itself near one of the arms of our galaxy, and as a result, we are presented with an excellent view of the bright nebulae that adorn the arm. These include the Eta Carina and Rosette nebulae and, closest at hand, the Orion Nebula. The Trifid, Lagoon and Eagle nebulae belong to another spiral arm nearer the center of the galaxy than our own.

The stars that illuminate the Orion Nebula are celestial infants, some of them less than five hundred thousand years old. One is estimated to have begun shining only about two thousand years ago; still embedded in the dark cloud from which it formed, it is invisible in these photographs but can be detected at the wavelengths of infrared light, which penetrates the dust and gas as visible light cannot.

---

5 The Horsehead Nebula is part of a large dark cloud of which the Orion Nebula forms a bright spot. Here, a little less than one hundred light years from the Orion Nebula, part of the enormous dark cloud may be seen silhouetted against a slightly more distant part of the cloud whose sheets of gas have been excited by starlight until they glow. The Horsehead itself is an eddy, a slowly spinning ball of gas that may be expected eventually to condense into new stars. Swirling at velocities approaching fifteen miles per second, it will one day no longer resemble a horsehead, having transformed itself like an earthly cloud on a summer day.

6 (overleaf left) Astronomers studying the Orion Nebula (= M42 = NGC1976) have found evidence of infant stars there, still wrapped in swaddling clothes of gas and dust (page 30).

7 (overleaf right) The interior of the Orion Nebula, known as the Trapezium, glows with the delicate green hue of ionized oxygen. Although the nebula is composed primarily of hydrogen gas, molecules of oxygen and formaldehyde have been found scattered through it, as have many others including those of carbon monoxide, hydrogen cyanide, ammonia, water and methyl alcohol. It is not yet understood just how the atoms of these thin clouds managed to link up into such complicated molecules.

THE EAGLE NEBULA

If we could speed up our sense of time until thousands of years were speeding by in the wink of an eye, we would see bright nebulae like these burst into light, deliver themselves of a shower of stars, then fade back into darkness. As it is, we see each nebula frozen at a stage in the process. The light that sets the nebula aglow comes from bright young stars that recently formed within it. The dark congealings of gas and dust, particularly prominent in the Eagle Nebula, are on their way to forming more stars.

The study of what might be called stellar embryology is still in its early stages, and a great deal remains to be learned about how galaxies make stars. But the story can be recounted at least tentatively in general terms.

The interstellar clouds arrayed through the vast tracts of space in a galaxy like ours abide most of the time in a state of passivity. They toil sluggishly along in space, responsive to the ghostly promptings of the galaxy's magnetic and gravitational fields. Once in a long while a star happens by, gobbling up a swath of the cloud as it goes. Nothing much else happens. At any given time much of the interstellar medium is a heavenly Dead Sea.

But through this sea pass waves. Density waves resulting from resonances generated by the gravitational interaction of the stars of the galaxy propagate across the galactic disk in a spiral pattern. When a density wave passes through an interstellar cloud, its effect is to compress the cloud. If the cloud is terribly thin, the passage of the wave will have only a temporary effect, fleeting as the stirring of dead leaves in a breeze. But where the interstellar cloud was sufficiently dense to start with, the wave can compress it until the cloud's gravitational field becomes strong enough to begin to draw the cloud still more tightly together. Once begun, this process will tend to continue. The cloud forms knots and eddies of ever-increasing density that draw more of the surrounding gas and dust in upon themselves, growing to become globules like the Horsehead Nebula and those that we see in the Eagle Nebula. Squeezed in the grip of their own gravity, the globules grow still more dense. Their interiors heat as the density increases. Ultimately, they become sufficiently hot and dense for nuclear fusion, the mechanism that powers hydrogen bombs, to begin at their centers. Light and heat pour forth. A star is born.

Meanwhile, the density wave proceeds on its way, leaving new stars scattered behind it, like a planter broadcasting seed.

---

8 The Eagle Nebula (= M16 = NGC6611), a glowing cloud of gas, measures some seventy light-years in diameter; on the scale of this photograph our solar system would be a microscopic speck.

10 A giant star-making machine, the Rosette Nebula wreathes some of the young stars that it created.

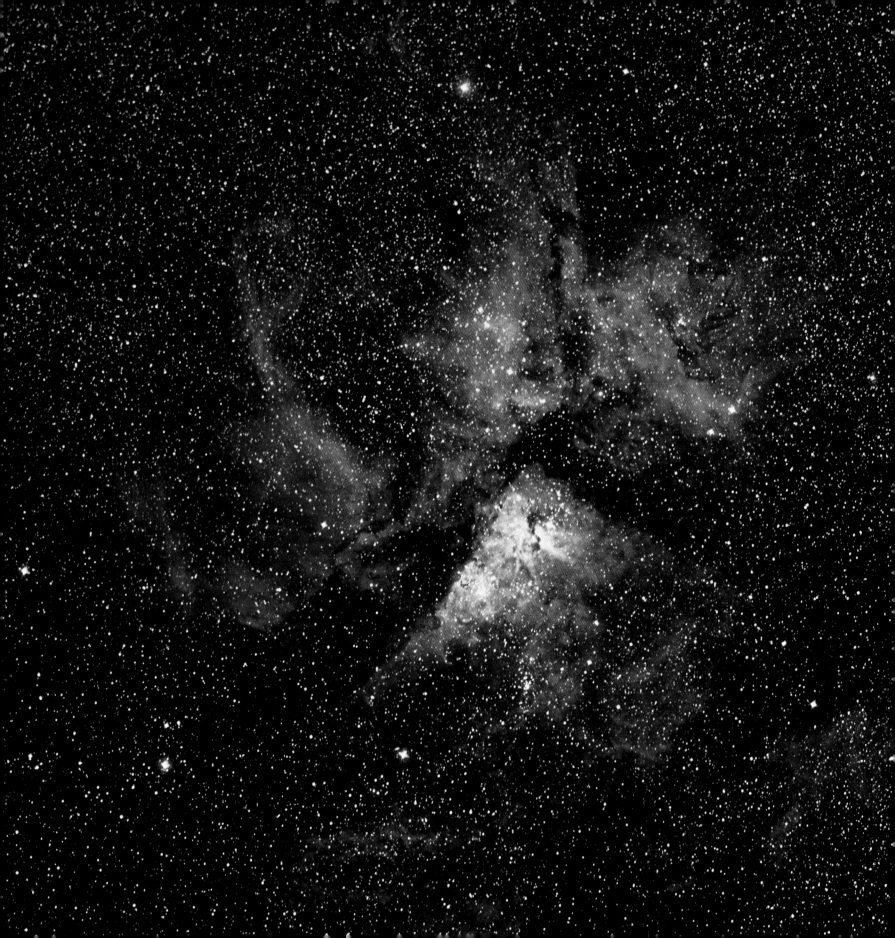

THE ETA CARINA NEBULA

As the sun pursues its orbit around the center of the Milky Way Galaxy, from time to time it passes through one of the galaxy's spiral arms. At the present epoch, when we humans have come upon the scene, built telescopes, and taken an interest in these matters, the sun and its planets lie near the inner edge of one of the arms. We have named it the Orion arm, after the bright constellation most of whose stars lie within it.

When we look in toward the center of our galaxy, we are confronted by the long flank of the next spiral arm in from the sun, which we have named the Sagittarius arm. Here again we see an array of bright nebulae, stretched out before us like the lights of a passing ocean liner. Among them are the Trifid, Lagoon and Eagle nebulae.

These glowing clouds are bright spots in an archipelago of dark gas and dust ranged along the length of the arm. The sinuous black veins winding through the star field in the photograph of the Eta Carina Nebula are not empty space, but are part of a meandering dark cloud to which the bright nebula belongs. The bright nebula stems from the dark cloud, like a blossom on a black branch.

---

11 The Eta Carina Nebula (= NGC3372), like most bright nebulae, is an illuminated portion of a larger dark cloud, seen here silhouetted against the background stars.

12 A shock wave may be seen moving like a storm front through the interstellar medium in this view of the interior of the Eta Carina Nebula.

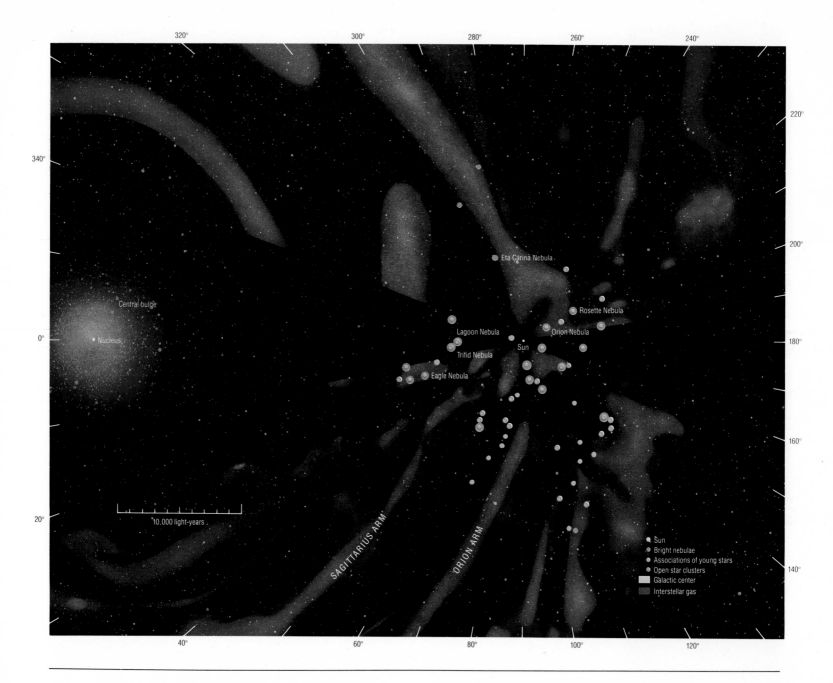

Figure 3. The Sun's Place Among the Spiral Arms of Our Galaxy
Beauty increases with proximity, says an informal maxim in astronomy, and as this map shows the beautiful bright nebulae whose photographs appear in this section belong to spiral arms of our galaxy that lie close to our solar system. The Lagoon, Trifid and Eagle nebulae are associated with the Sagittarius Arm, while the Eta Carina, Orion and Rosette nebulae are associated with the Orion Arm. The sun lies between these two arms, close to the inner edge of the Orion Arm. The cone-shaped zones to the left and right of the sun that interrupt the contours of the arms are not real, but indicate areas where intervening dark gas clouds block our view and prevent accurate mapping of the areas beyond. Like early maps of the New World, this map should be considered only a rough approximation; we are just beginning to learn our way around our own galaxy.

THE TRIFID AND LAGOON NEBULAE

These nebulae are three-dimensional structures. They have considerable depth as well as their more evident articulation in the two other dimensions. This is especially well illustrated by the Trifid Nebula. Starlight and glowing gas within light it up like a ship's lantern; the dark portions of the cloud that seem to divide it into thirds (hence the name "Trifid") are foreground parts of the cloud that we see in silhouette like struts in a lantern's globe. The red hues are produced by glowing hydrogen gas. The ice-blue regions are primarily dust particles in the cloud reflecting the light from stars in the nebula. These young stars, very hot, radiate generously in the energetic wave lengths of blue light.

The Lagoon Nebula was named by an observer who fancied that the dark rift running across its face resembled a map of a harbor. Actually this feature is almost certainly a dark foreground portion of the cloud, like the "struts" that cut across the Trifid Nebula.

Young stars abound in the Lagoon Nebula, many of them still flaring up erratically as they struggle to bring their gravitational and radiative forces into the balance that will be required for each to settle down into a tenure as a stable star. The intense light from these young stars and from the glowing rebula enshrouding them makes them some of the brightest objects in the galaxy.

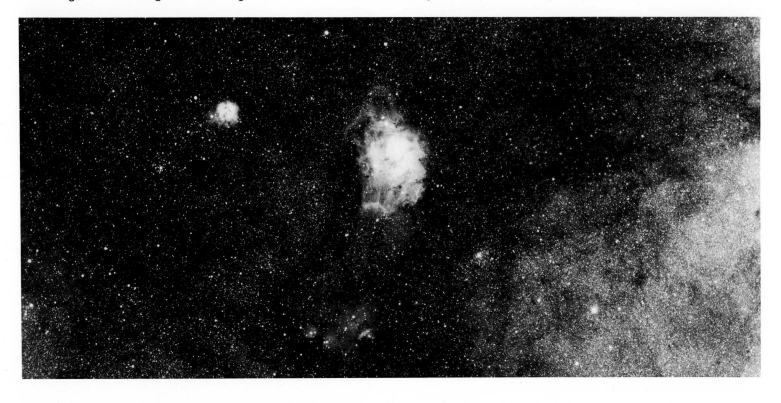

13 A field view of the Trifid and Lagoon nebulae suggests that both are tangled in the same large and otherwise dark interstellar cloud. A final determination of this awaits a better assessment of the distance separating the two nebulae, currently estimated to be as much as five thousand light-years or more. The Trifid, according to several estimates, lies considerably farther away from earth than does the Lagoon.

14 **(overleaf left)** The dark globules of the Lagoon Nebula (= M8 = NGC6523), hatcheries of young stars, are estimated at this stage in their collapse to measure a light-month or two in diameter.

15 **(overleaf right)** The Trifid Nebula (= M20 = NGC6514) belongs, like its neighbor bright nebula the Lagoon, to the Sagittarius arm of our galaxy.

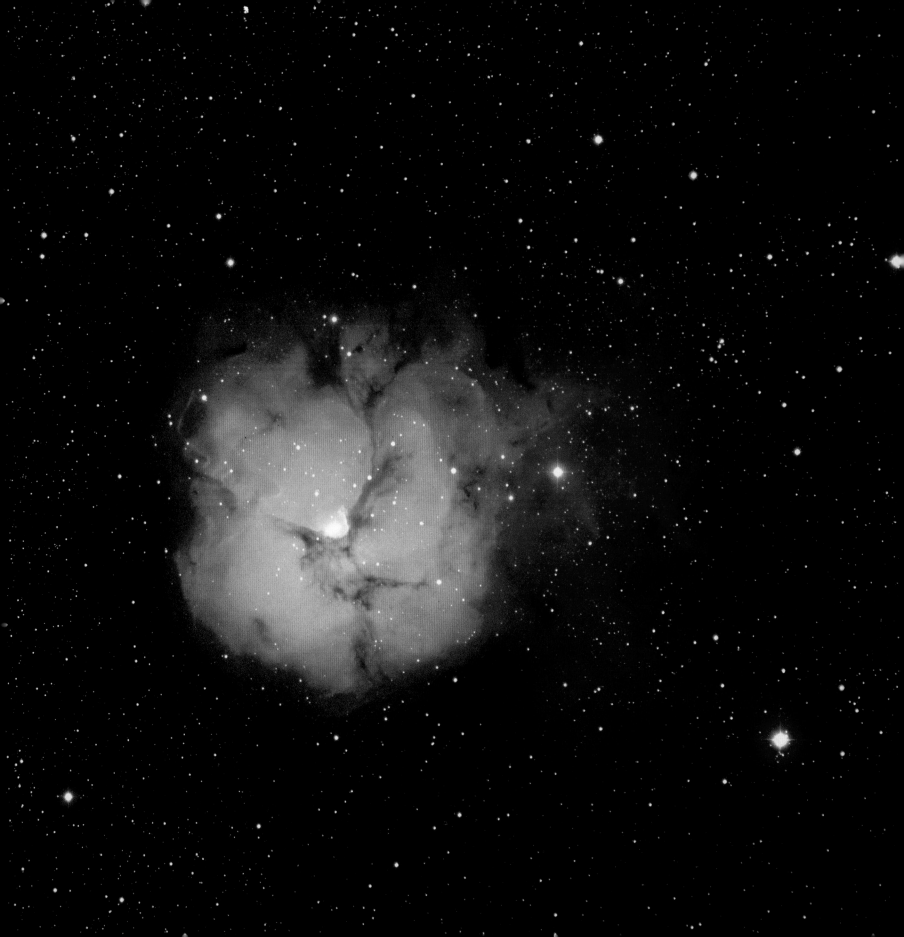

OPEN STAR CLUSTERS

Stars that were born together stay together for a while, in associations known as star clusters. There are two distinct kinds of star clusters. Open clusters, like those seen on these pages, range in population from a few dozen to a few hundred stars. Globular clusters, like those seen on the following pages, are much larger and have populations that range into the millions of stars. Globular clusters are gravitationally fairly stable, almost like tiny galaxies all their own, and many are quite old. Open clusters, less permanent, typically are young. The stars of the Pleiades (page 49) are not much more than one hundred million years old.

The reason most open star clusters are young is that they cannot hold together for long. Lacking the imposing mutual gravitation of the hundreds of thousands of stars found in a globular cluster, they tend to fall apart as time goes by, losing their member stars by way of both internal and external influences. Stars on the outskirts of the cluster may be lost to the gravitational tug of a passing star, or of another star cluster, or of the galaxy as a whole. More often, stars are lost internally, when a lesser star in the cluster passes near one of its more massive compatriots; the gravity of the massive star accelerates the lesser one, like a skater in a crack-the-whip, hurling it out of the cluster.

The loss of each star reduces the gravitational potential of the cluster as a whole, leaving it increasingly vulnerable to further defections. Ultimately the cluster is reduced to only a skeleton crew and may dissipate entirely into the general galactic population. Most of the stars we see in the galaxy today, even solitary stars like our sun, may once have belonged to open clusters.

---

16 The young cluster NGC3293 lies about ten thousand light-years from earth.

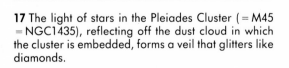

17 The light of stars in the Pleiades Cluster (= M45 = NGC1435), reflecting off the dust cloud in which the cluster is embedded, forms a veil that glitters like diamonds.

The more than one hundred globular star clusters associated with our galaxy have two particularly intriguing characteristics. One is their age. Globular clusters typically are very old. Some have been estimated to be over fifteen billion years old, a longevity comparable to that of the galaxy itself. The other noteworthy characteristic of the globular clusters is their distribution in space. While most of the bright stars of our galaxy are situated along the plane of its disk, globular clusters are found above and below the plane as well.

These two attributes of globular clusters are of great interest to researchers concerned with how the galaxy may have formed. Most of these theorists have determined, by reconstructing the physics of the situation, that our galaxy began as a more or less spherical aggregation of gas that subsequently collapsed to form the disk it describes today. Opinions differ as to how long such a collapse took to occur, and as to how many stars had formed from the primordial gas by the time of collapse, but most agree that long ago the Milky Way protogalaxy was roughly spherical in shape. This impression gains credence from the fact that the globular clusters, made up of very old stars, are still found well out of the galactic plane, distributed in a spherical volume of space that may replicate the dimensions of the old Milky Way protogalaxy. The distribution of globular clusters may constitute a relic of the way the galaxy once was, like the charred frame of a building whose walls have collapsed in a fire.

The clusters themselves are imposing. The largest globulars have millions of stars and, at least in the case of those that lie well away from their parent galaxy, it is difficult to decide just where to draw the line distinguishing a large globular cluster from a dwarf galaxy.

Dramatic perspectives on the cosmos are implied by the location of many globular clusters and by their wealth of stars. Imagine, for instance, that the sun and earth were situated not here in the plane of our galaxy, but in the outer fringes of a remote globular cluster that lies well away from the plane. For a half of each year our night skies would be jammed with the brilliant stars of our home cluster, their light so intense that darkness would never really fall. For the other half of the year, during the seasons that found us on the side of the sun away from the globular cluster and toward the Milky Way Galaxy, we would see the galaxy laid out flat from one horizon to the other.

A price paid by remote globular clusters for their enviable

18 The light from old red giant stars warms the hues of the globular cluster NGC2808, twenty-five thousand light-years away.

19 The globular cluster M13 (= NGC6205) is approximately two hundred light-years in diameter, but most of its more than one million stars occupy a central region whose diameter is under one hundred light-years; in these relatively crowded quarters the average clearance is one star per cubic light-year.

view of the cosmos is sterility. The interstellar gas and dust needed to make new stars is concentrated along the galactic plane. Globular clusters lying far from the plane are denied these riches, and so stars rarely form there. The stars we see in globular clusters are mostly old survivors born long ago. The virtues of these gerontocracies are those of the baroque; they are refined and enduring, but in terms of stellar evolution they represent nearly the end of the line.

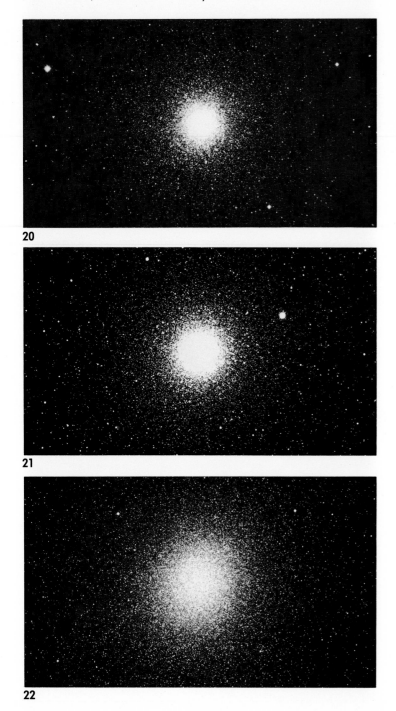

20

21

22

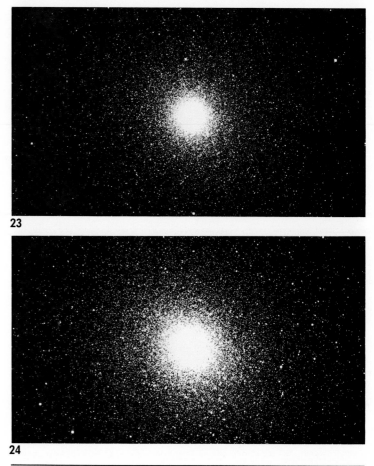

23

24

20–24 These globular clusters are: **20**: M3 (= NGC5272), a large old cluster with an estimated five hundred thousand stars, many of them dwarves and other degenerates that have seen better days; **21**: M15 (= NGC7078), almost fifty thousand light-years distant and well away from the plane of our galaxy; **22**: Omega Centauri (= NGC5139), the brightest known globular cluster, which lies near the plane of the galaxy halfway in from the sun toward its center; **23, 24**: M5 (= NGC5904) and 47 Tucana (= NGC104), which both show evidence of flattening perhaps due to rotation.

25

25, 26 The globular clusters NGC6522 and NGC6528 (above) lie within the plane of our galaxy, passing their time in the company of billions of noncluster stars. The remote globular cluster NGC2419 (right), on the other hand, has wandered more than three hundred thousand light-years from the center of our galaxy, to a point near the limits of the galaxy's gravitational domain; this "intergalactic tramp" cluster follows a lazy orbit around the galaxy that will require over three billion years to complete.

26

The Death of Stars

The lesson of life that nothing lasts forever is echoed in the death of stars. Though symbols of constancy, stars ultimately must perish. And, as we find elsewhere in nature, the mortality of the individual plays a part in the evolution of the whole.

Stars die in accordance with how they lived. Modest stars like our sun conclude their careers modestly. Having depleted most of their fuel, they expand to become blowsy, dimly glowing giants that shrug off their outer atmospheres and settle down into retirement as dwarf stars. More massive stars conclude their brilliant reign in a more spectacular fashion—they explode. Extraordinarily massive stars wind up their extravagant careers by exploding with extraordinary force.

This suitability of death to life in stars pertains also in the design of stellar tombs. The remains of ordinary stars take the form of inconspicuous dwarves. More massive stars collapse to form more noticeable monuments, the neutron stars, whirling stellar cinders compressed to a state harder than diamond. The most imposing stars may collapse so forcefully as to form black holes, cutting themselves off from the rest of the universe and realizing the ambitions of the Pharaohs in that their remains can never be exhumed.

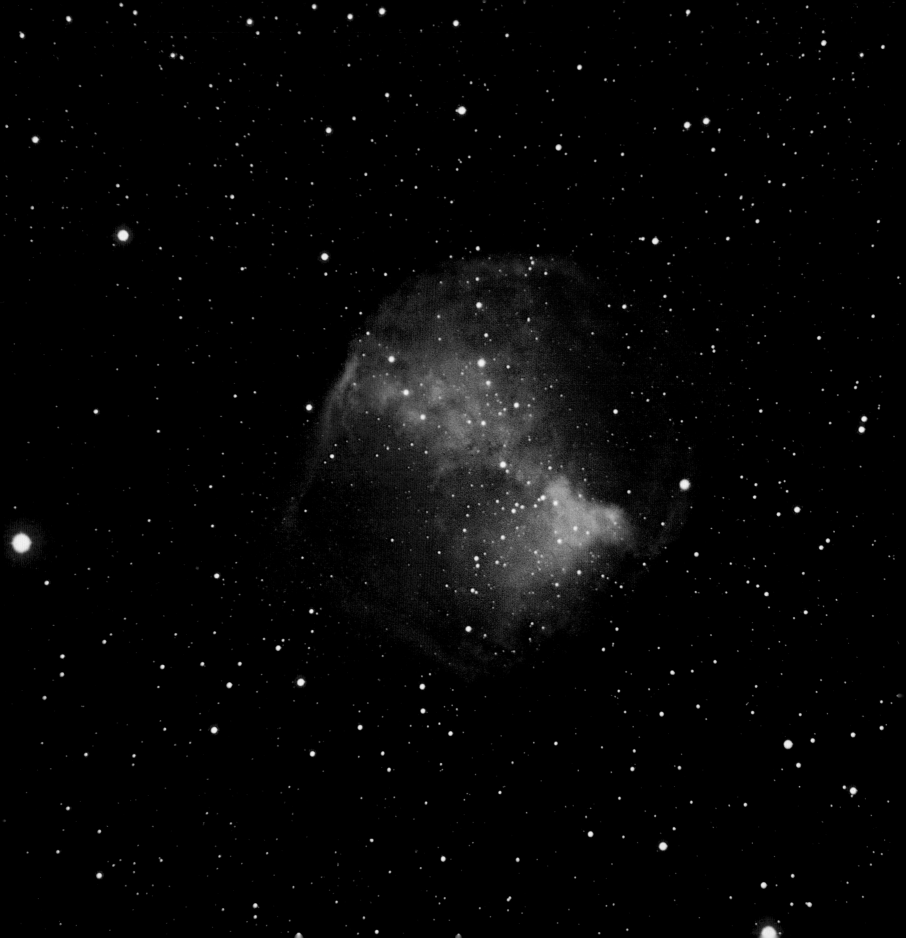

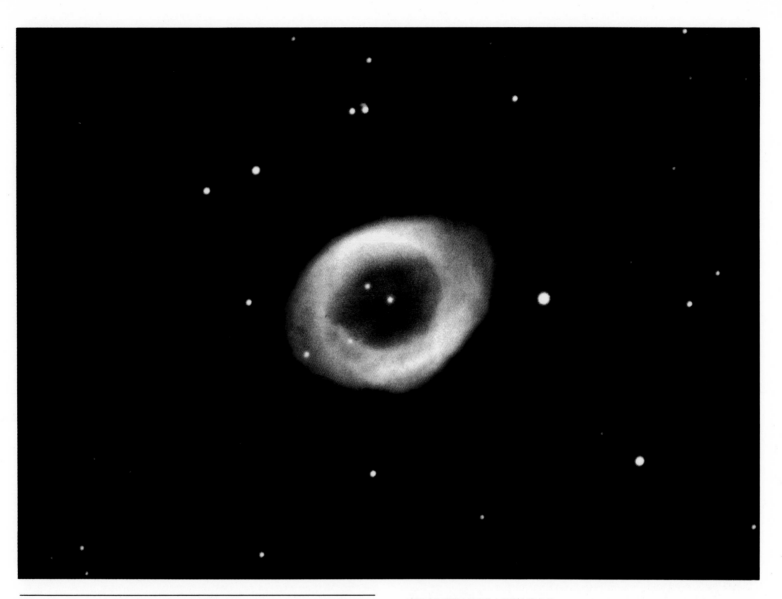

27 The "planetary" nebula M27 (=NGC6853) consists of an envelope of gas ejected from the central star about fifty thousand years ago; expanding at a rate of seventeen miles per second, the gas bubble has achieved a diameter of over two light-years.

28 The "planetary" nebula M57 (=NGC6720), known as the Ring Nebula, consists of gas ejected from the central star some twenty thousand years ago; its rainbow hues result from the glow of various excited atoms in the cloud, among them those of hydrogen, helium, oxygen, nitrogen, sulphur, and—most familiar among the fluorescent elements here on earth—neon (above).

"PLANETARY" NEBULAE

Stars cannot harbor their substance forever, but must return a substantial portion of it to the interstellar medium from which they formed. They do this to some extent all the time; the sun ventilates a steady stream of particles into space in the form of the "solar wind," and some giant stars produce gale-force stellar winds. But it is when they approach the end of their careers that stars disgorge themselves dramatically. This occurs when a star finally exhausts the fuel—mostly hydrogen gas—that has enabled it to keep burning. The process may be described in a radically simplified fashion by saying that the core, its fuel spent, cools and collapses into a dwarf remnant while the massive outer envelope of the star—the part that had surrounded the core—boils away into space under the impulse of its own heat.

The so-called "planetary" nebulae like these are stars that we have caught in the act of discharging such a shell of gas. What appears to be a ring surrounding the star is actually a thick shell or bubble. As the shell expands into space, light from the degenerate star it left behind excites the gas to glow. The delicate colors of the shell result from the excitation of its various gases, among them hydrogen, oxygen and nitrogen. The term "planetary" is another of those misnomers that complicate astronomical lexicography; it results from an error by early astronomers, who, equipped with only small telescopes, noticed a vague resemblance between these shells of gas and the disks of the planets of our solar system.

"Planetary" nebulae are passing things. Each will go on expanding until it has dissipated into interstellar space. The "planetary" nebulae seen in these photographs are typically only a few tens of thousands of years old; in another few tens of thousands of years they will have all but vanished. Meanwhile, the core of the star left behind will in most cases have settled down as a dwarf, able to glow dimly for a very long time before itself fading away. The "planetary" nebula phase therefore represents a brief episode late in the life of a star, one that costs it no greater portion of its time than a nonfatal heart attack deducts from the life of a human being.

Each year in the Milky Way, a few aging stars slough off their skin in this manner. The result is that thousands of "planetary" nebulae ornament our galaxy at any given time. In all, they add enough material to the interstellar medium each year to form five new stars the mass of the sun.

ERUPTING AND EXPLODING STARS

Novae and supernovae are stars that divest themselves of material not with the orderly demeanor of a "planetary" nebula but with explosive violence. The name "nova," meaning new star, is an index to the spectacle they create; exploding as a nova or supernova, a previously nondescript star can flare up so brightly that it dominates the sky, creating the illusion that a new star has come into being where none was before.

The terms "novae" and "supernovae," however, cover a considerable range of stellar violence; a very minor nova may not be much more traumatic to the star involved than is the ejection of gas to form a "planetary" nebula, while supernovae may attain apocalyptic levels of violence. When the mechanisms of stellar explosions are better understood, it may be possible to assign supernovae, novae and the events that produce "planetary" nebulae places within a single spectrum describing the means by which stars divest themselves of excess baggage on the way to their graves.

Most spectacular of these stellar death rites are the supernovae. Their luminosity may surpass that of billions of normal stars, and they can eject enough material to make several of our solar systems. Such an event would spell disaster (literally, for the word "disaster" comes from the Latin for "unfavorable aspect of a star") for anyone having the misfortune to live in the celestial neighborhood of a supernova. Planets of the exploding star would be vaporized, and planets of other stars lying within a few light-years would be bathed in a quantity of radiation sufficient to sterilize them. Yet, as we find with terrestrial disasters like forest fires, earthquakes and hurricanes, the destructive forces of nature contain seeds of creation. Exploding stars play a vital role in the ecology of the galaxy.

Supernovae build heavy elements. All stars are engaged to some degree in evolving complicated atoms out of the simple atoms of hydrogen and helium. Generally speaking, the more massive a star and the hotter its interior, the heavier the elements it can construct there. Very massive stars are capable of forging atoms as heavy as those of iron. But they are unable to go further. Iron atoms, highly stable, cannot be broken down and rebuilt into heavier atoms even in the interior of a hot star. To surmount this barrier calls for a drastic solution, and a supernova is just that. In its intense heat a wide array of heavy atoms are forged, then sprayed into interstellar space.

So supernovae are not solely celestial death spasms, but represent the crowning achievement of a star that caps a lifetime of element-building by seeding its cosmic neighborhood with atoms as exotic as those of uranium and gold. Stars and planets that form subsequent to the supernova inherit these heavy elements as part of the interstellar cloud from which they are made. That is how we inherited the heavy elements we find here on earth: They had been injected into the interstellar medium by stars that exploded before the sun and earth condensed from that medium.

It is curious to reflect that the materials employed to build the telescopes that took the photographs on these pages, and the carbon used to ink these words, were created within

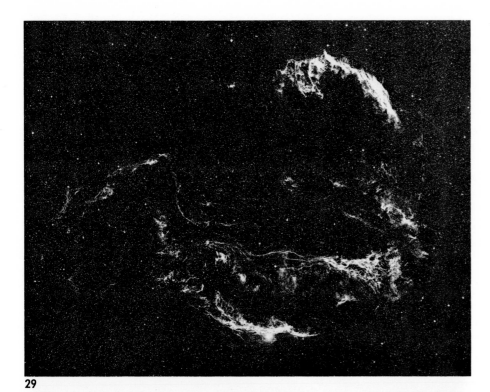

29 The Veil Nebula (=NGC6960/6992), ejecta from a star that exploded thirty or forty thousand years ago, is now nearly one hundred light-years in diameter. It continues to swell, a slowly bursting bubble, its rate of expansion decreasing as it entangles itself with other interstellar clouds. Rich in heavy elements, it seeds the interstellar medium like a gigantic thistledown.

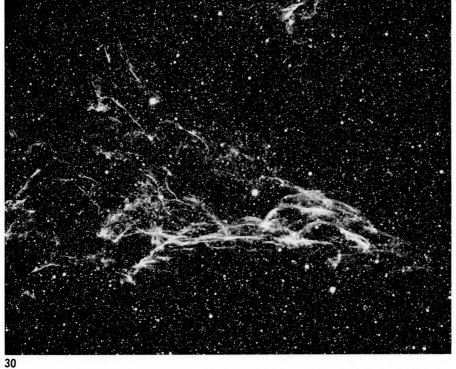

30 A few of the streamers of the aptly named Veil Nebula are seen here in detail; the segment visible in this photograph measures approximately twenty light-years across.

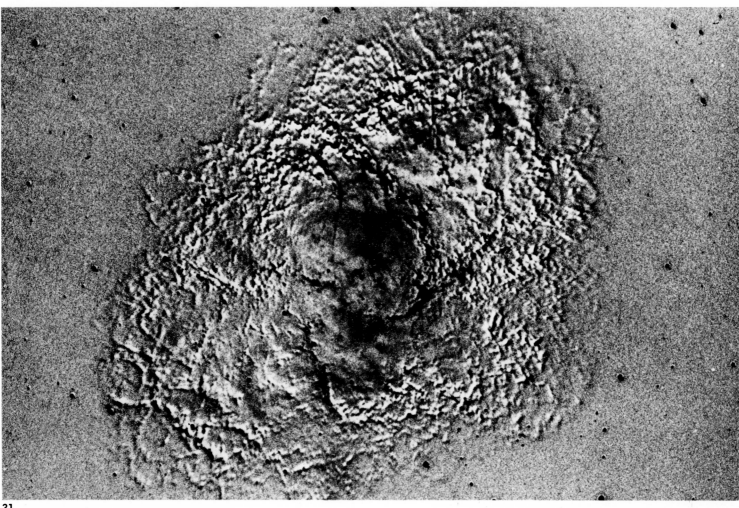

31

stars like those we now contemplate. The stars facilitate our study of them.

When a star explodes it leaves behind two sorts of debris —that which it has blown into space and that which stays behind. The material blasted into space expands until it has dissipated and mingled with the rest of the interstellar medium. The material left behind collapses to form a dwarf star, or a denser object known as a neutron star, or an object so dense that it swallows up its own light and becomes a black hole.

The Crab Nebula (right) is the remnant of a supernova that occurred only five thousand light-years from the sun, quite recently in cosmic time; its light reached earth in July 1054 and was noted by Chinese, Arab and American Indian observers. Thanks to its proximity, the Crab Nebula can be studied in detail. It provides us, so to speak, with a warm body on which to conduct a supernova postmortem.

The crushed cinder at its center, all that remains of the star,

31 Expanding at a rate of six hundred miles per second, the Crab Nebula supernova remnant can be seen to grow in size as the years go by. Here a positive print, in which the nebula appears light in color, has been superimposed on a negative print made fourteen years later, in 1964. Expansion of the cloud during those fourteen years has offset the dark from the light images, producing a bas-relief effect.

32 Wreckage of an exploded star, the Crab Nebula (=M1=NGC1952) consists of an envelope of material blasted into space when the star blew up, and at its center a collapsed dwarf, or neutron, star.

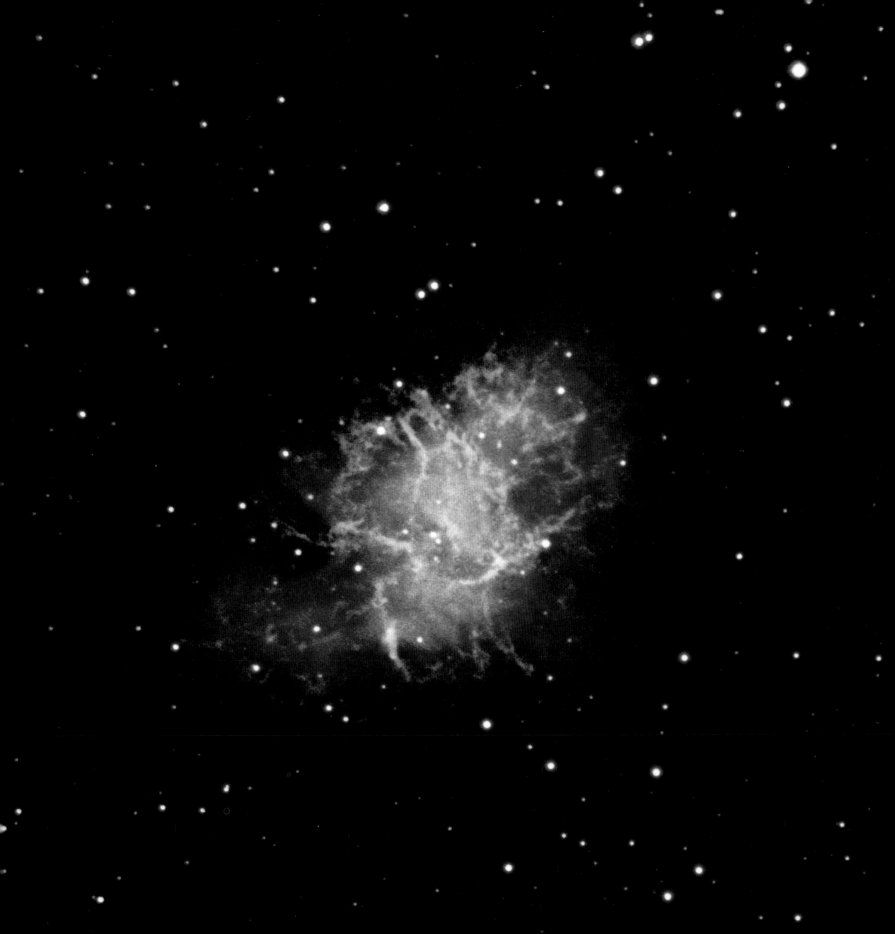

is compressed to such a density that a spoonful of it would weigh millions of tons. Spinning thirty times per second, this strange object emits pulses of energy in both radio and optical wavelengths; astronomers call it a neutron star, since it consists primarily of neutrons packed closely together, or a pulsar, when referring to the pulses of radio waves it emits.

The surrounding envelope of material hurled off in the explosion glows with excitation produced by radiation pouring out from the cinder. The colors of its filaments come principally from ionized hydrogen, carbon and sulphur.

The effects of a supernova like the one that created the Crab Nebula are varied and subtle. Some are intimate; high-energy particles of the sort we call cosmic rays are produced by the remnant and some of these strike the earth, where they are capable of shattering genetic material in reproductive cells and creating mutations that alter, however slightly, the course of the evolution of life here. No less important in the history of humankind are the intellectual effects produced by the sight of a supernova, as when the Danish astronomer Tycho Brahe saw a supernova in 1572 as a young man in his twenties and was emboldened to question the philosophical authority of Aristotle, who had held the realm of the stars to be eternally unchanging.

BLACK HOLES

The cosmos is so abundant in things we can see—stars, bright nebulae, the planets here in our solar system—that we may neglect to consider how much there is in the cosmos that we do not see. A considerable portion of the mass of the universe, perhaps most of it, is inconspicuous or even invisible.

Most of the inconspicuous constituents of the cosmos ought to be detectable sooner or later, given good enough equipment. Planets of other stars, undetectable at present, ought to be discernible with specially designed telescopes operating in space. Dwarf stars too dim to be photographed with existing telescopes can be expected to evidence themselves when telescopes with far greater light-gathering power have been built. Matter in highly rarified forms, such as the ethereal clouds of cold hydrogen gas found in intergalactic space, can be located by charting energy emitted by their scattered atoms at the invisible wavelengths of radio or X-rays.

But there is one form that matter may take that renders it genuinely invisible, hides it away from view forever—a black hole.

The term "black hole" describes the outward appearance of a class of astronomical objects that are compressed to such a high density that their gravitational field prevents even their own light from escaping them. A black hole has in effect retired from the outside world. No energy of any sort emerges from it. A black hole offers us no picture of itself, and if we tried to take a flash photograph of it, it would simply swallow the light from the flash. Hence the name.

Theoretical astrophysicists envision several ways that black holes might come into being. One is by way of the collapse of the core of a star. We have seen that massive stars that have exploded as supernovae leave behind degenerate cores that, caught in the grip of their own gravitation and no longer propped up by the energy generated by nuclear fusion, can collapse to form the highly condensed objects known as neutron stars. But if the core left behind after a supernova is sufficiently massive (more than about three times the mass of the sun), it could collapse right through the neutron star stage to become so small and gravitationally powerful that it swallows up the light it produces.

Another possibility is that black holes might be formed by the collapse of still more massive objects, such as the nuclei of galaxies or of large globular star clusters. Or one can conceive of mini-black holes, tiny as subatomic particles.

These are the theoretical possibilities. Whether black holes exist, and in what abundance, depends upon the extent to which nature has pursued the paths theoretically available to her. Our experience here on earth indicates that nature does pursue many exotic paths (e.g., bower birds), while declining to pursue others (e.g., unicorns), and by extrapolation beyond the earth we can imagine an enormously heightened variety (e.g., extraterrestrial unicorns?).

If massive stars do collapse into black holes, then we can estimate that there are some one hundred million black holes in the Milky Way today, each with a mass of at least several times that of our sun, each the remnant of a huge star that exploded in the past. If, in addition, the nuclei of many galaxies harbor black holes, we would expect each to weigh tens of thousands of times the mass of the sun, owing to the abundant opportunities for these black holes to gobble up interstellar gas and other material available to them in the crowded central galactic regions. And if, to take the extreme case, those scientists who envision ubiquitous subatomic black holes are correct, more than ninety percent of the mass of the universe might be cached away in the realm of the invisible, demoting the visible cosmos to the status of something of an afterthought. It remains for future observation to determine where the truth lies along the spectrum ranging from a universe with no black holes in it to a universe composed mostly of black holes.

To search for a black hole is to embark upon a quest of the sort that would have delighted Lewis Carroll. One is looking for the invisible. How to go about it?

A promising strategy is to seek evidence of the effect of black holes upon their surroundings. This might be called the Invisible Man Approach, after the character in the H. G. Wells science fiction novel who is safe so long as he does nothing to interact with his environment, but risks detection if he blunders into someone or tries to make off with a sandwich. Black holes, infinitely hungry, will consume any material that comes near enough for them to trap it. A black hole embedded in an interstellar cloud will swallow portions of that cloud. A black hole in a double star system will, if it gets close enough to its companion star, strip material from the companion and consume it. A black hole feeding ground ought to be detectable by observing energy, notably X-rays, thrown off by the doomed material as it swirls down into the black hole.

Orbiting X-ray telescopes have succeeded in locating several X-ray sources that very possibly represent black holes. The first such candidate to be discovered was Cygnus X-1. Here the black hole, if it is such, is part of a double star system something over six thousand light-years away. The visible star in the system has a mass of thirty times the sun's, while its companion, alleged to be a black hole, has about six times the sun's mass.

Other powerful X-ray sources have been located at sites where we might expect to find black holes. These include the nuclei of some galaxies and of massive globular clusters.

The "event horizon" of a black hole, the boundary from within which nothing can escape, appears to be inviolable. This may be why black holes are so provocative to the human imagination, for few if any absolute boundaries have yet been found in nature. Human history is dotted with stories of the surmounting of them—the edge of the world, the speed of sound, travel in space—and an immediate human response to being told that no one can cross the event horizon of a black hole and return is to try to find a way to beat the system.

Structure of the Milky Way

As we survey the stars and interstellar clouds surrounding us, we see that they are organized not randomly but in accordance with a pattern. Most are incorporated into a flattened disk that we now know to be the disk of our galaxy. Its appearance from our vantage point is that of a broad river of light stretching across the sky and glowing with the combined light of myriad stars—the Milky Way. Once we learned that it was in fact our view of our own galaxy, we named the galaxy after it.

The photographs on the following pages show portions of the Milky Way that lie toward the center of our galaxy. Our sun is more than halfway out from the center, so the richest star fields from our vantage point are those we see when we look back toward the center rather than out away from it.

Our appreciation of the view is enhanced if we strive to imagine it in three dimensions. The interstellar clouds display clear evidence of dimensionality; their loops and festoons can be seen to have depth, like the curving ribs of a skeleton. The concentration of interstellar material along the plane of the galaxy is typical of spiral galaxies; compare this view with that of the external galaxy (NGC2683) seen edge-on (page 111).

---

33 (overleaf left) Rich fields of stars and interstellar gas and dust characterize the Milky Way in the direction of the constellation Cygnus. Here is to be found an enormous gas bubble apparently produced by a series of exploding stars, and the X-ray source Cygnus X-1, thought to be a black hole.

34 (overleaf right) The Milky Way. When we look toward the center of our galaxy, we see hundreds of millions of stars that, like our sun, lie along the galaxy's flattened plane. These portions of the Milky Way are found in the southern skies of earth in the direction of the constellation Sagittarius.

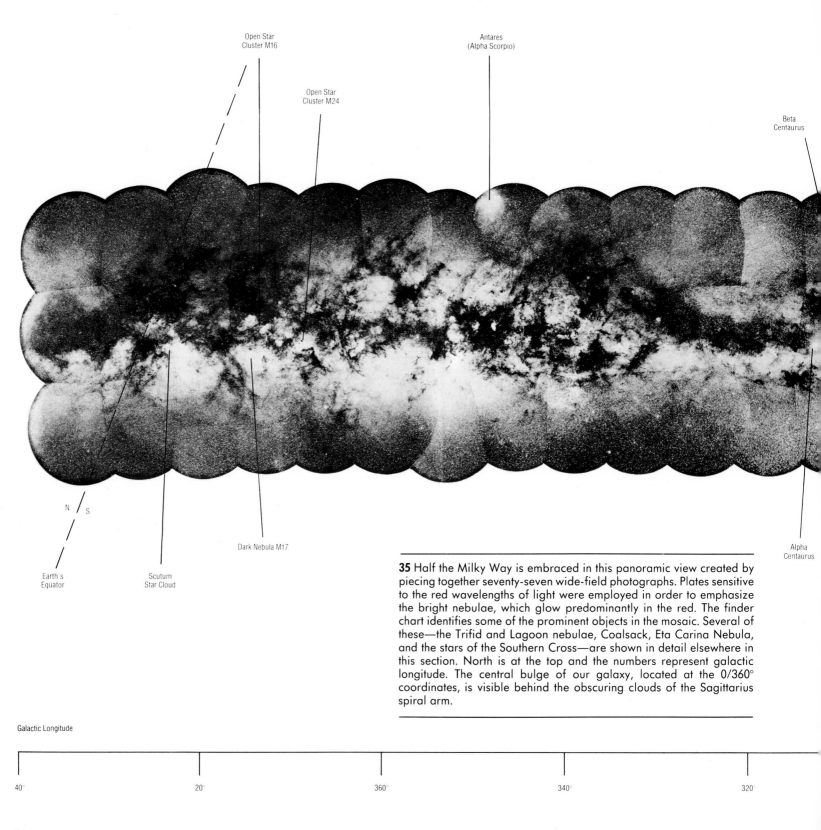

Open Star
Cluster M16

Open Star
Cluster M24

Antares
(Alpha Scorpio)

Beta
Centaurus

N / S

Dark Nebula M17

Alpha
Centaurus

Earth's
Equator

Scutum
Star Cloud

35 Half the Milky Way is embraced in this panoramic view created by piecing together seventy-seven wide-field photographs. Plates sensitive to the red wavelengths of light were employed in order to emphasize the bright nebulae, which glow predominantly in the red. The finder chart identifies some of the prominent objects in the mosaic. Several of these—the Trifid and Lagoon nebulae, Coalsack, Eta Carina Nebula, and the stars of the Southern Cross—are shown in detail elsewhere in this section. North is at the top and the numbers represent galactic longitude. The central bulge of our galaxy, located at the 0/360° coordinates, is visible behind the obscuring clouds of the Sagittarius spiral arm.

Galactic Longitude

| 40° | 20° | 360° | 340° | 320° |

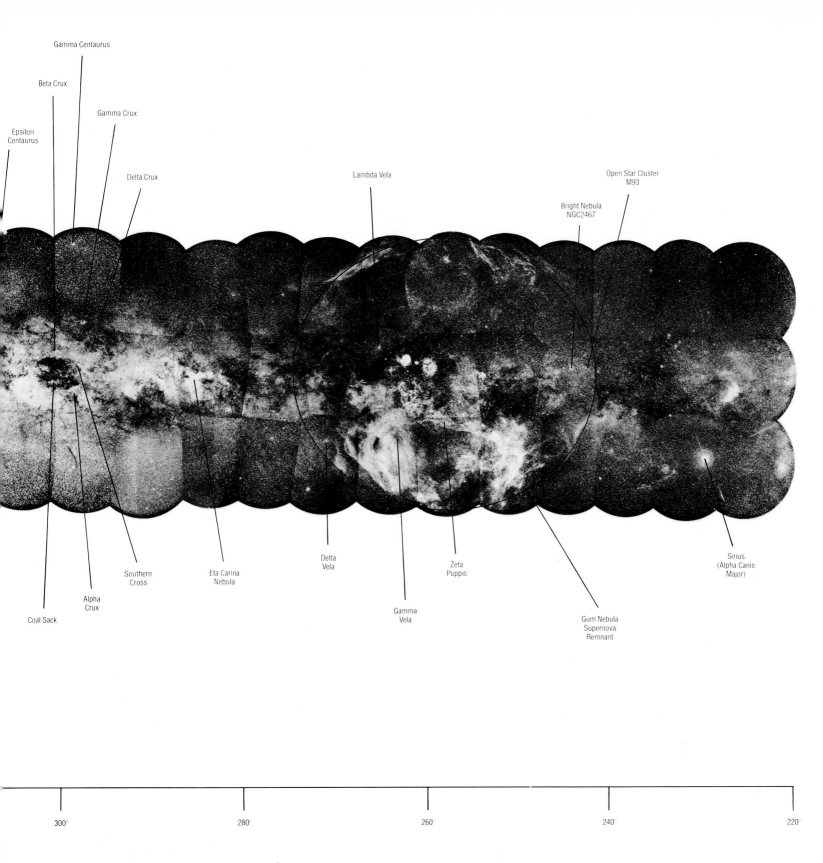

Epsilon
Centaurus

Gamma Centaurus

Beta Crux

Gamma Crux

Delta Crux

Lambda Vela

Open Star Cluster
M93

Bright Nebula
NGC2467

Coal Sack

Alpha
Crux

Southern
Cross

Eta Carina
Nebula

Delta
Vela

Gamma
Vela

Zeta
Puppis

Gum Nebula
Supernova
Remnant

Sirius
(Alpha Canis
Major)

300°　　　　　　　280°　　　　　　　260°　　　　　　　240°　　　　　　　220°

II/The Local Group of Galaxies

*He showed me a little thing,
the quantity of an hazel-nut,
in the palm of my hand;
and it was as round as a ball.
I looked thereupon with the eye
of my understanding, and thought:
What may this be? And it was
answered generally thus: It is
all that is made.*

—ST. JULIANA

A Journey Out of Our Galaxy

We speed out of the Milky Way in our imaginary spaceship like divers ascending from the depth of the sea. The myriad bright stars which had been our companions now diminish in number, then fall away behind us. In their stead we are left with the scattered stars of the galactic halo. Most are dim dwarves, remnants of stars that formed more than ten billion years ago, when the infant galaxy was more nearly spherical and had not yet collapsed to its present flattened shape. A few "runaways," younger stars ejected out of the plane of the galaxy by quirk gravitational encounters, flit among these elders like bright tropical fish venturing from their accustomed shallows.

To exit a galaxy is no handy matter, but eventually we attain a sufficient remove to be able to view the galaxy spread out below us. The central bulge of the galaxy looms directly under us, its shape and color like that of a hill of sand. The galactic disk surrounds it, a monumental tangle stretching to the celestial horizons. The glowing clouds of the spiral arms wend their way out through the disk, often obscured by intervening dark clouds, like a river cutting through a jungle. Here and there dark tattered towers are reared up out of the welter of the disk, masses of interstellar gas and dust that have been heaved out of the galactic plane in the course of the collisions of clouds and the explosions of stars.

We climb further and our view of the galactic disk improves. The time comes when we can discern the sun, a little yellow star nestled in the embrace of the outer reaches of one of the spiral arms, a dot of light barely visible through the ship's telescope. Here long ago was our home.

The globular star clusters make for spectacles close at hand. From time to time one of these chandeliers of stars passes abeam of our ship. We are tempted to stop and explore its hundreds of thousands of stars, but we are bound for territories more remote.

When the outermost of the globular clusters along our course has fallen away aft, we celebrate having left our home galaxy behind. The choice of demarcation is rather arbitrary, for we still lie well within the gravitational domain of our galaxy. But we feel the need of a cheering toast, for we are embarking upon the awful gulf of some of the emptiest space known in a universe that is mostly space. Our galaxy hangs behind us like a gong, its slowly diminished starlight painting shadows across our ship from aft, while ahead yawns the void, its only light the pearlescent background haze of a universe of galaxies.

Our eyes seek out landmarks lest we be seized by vertigo. Well off to port hangs the most evident galaxy in sight after the Milky Way, the Large Magellanic Cloud. Beyond it we can make out the less orderly patches of starlight that comprise the Small Magellanic Cloud and the Sculptor and Fornax dwarf galaxies. To starboard lie two other dwarves, the little Leo I and Leo II galaxies. We steer for them.

Seven hundred fifty thousand light-years separate the Milky Way from the Leo pair. Our activities during this phase of the voyage are those suitable to a long haul. We carve scrimshaw, repair our gear, read all the back issues of the National Geographic. Down below the stokers shovel whole planets' worth of fuel into the engines to maintain our acceleration. Leo I and II slowly grow in the forward viewports.

Now we can look back upon most of the Local Group in a single gaze.

The Milky Way Galaxy, though still imposing, has shrunk until it covers less than ten degrees of the sky; we can eclipse it with an outstretched hand. A little train of satellite galaxies stretches off to one side of the Milky Way. In the same part of the sky but far deeper in space hangs the spiral galaxy M33, and near it the majestic M31, dominant galaxy of the Local Group. Beyond them we can glimpse the elliptical galaxy Maffei I and its spiral companion Maffei II.

Will we in our little ship feel a last pang of leave-taking as we say goodbye to this corner of the universe, with its trillions of suns, in the light of one of which we came into being? Or has this already become too strange and remote to retain any of the warmth of home? We rocket past the Leo I dwarf galaxy and head out of the Local Group.

II/THE LOCAL GROUP

The Magellanic Clouds

The nearest galaxies to ours are the Magellanic Clouds. They are called "Magellanic" by virtue of their having been introduced to Western civilization by the crew of Ferdinand Magellan, whose circumnavigation of the earth took his ships beneath southern skies, where the clouds are to be seen. They are called "clouds" owing to their soft outlines and glowing appearance, which make them look something like scraps detached from the Milky Way. In the early decades of the twentieth century, a number of variable stars of the sort known as Cepheid variables, invaluable to astronomers as distance indicators, were identified in the Magellanic Clouds. Their discovery made it possible to establish that the clouds were too distant to be part of our galaxy, and had to be galaxies in their own right. The Large Magellanic Cloud is about one hundred fifty thousand light-years, the Small Magellanic Cloud about 250,000 light-years from the sun. Less than one hundred thousand light-years separates the two clouds.

The clouds lie well within the gravitational field of our galaxy and orbit it as satellites. This arrangement, small galaxies playing court to large ones, is commonplace in the universe; a major spiral like the Milky Way typically plays host to several satellites. In the case of the Milky Way, two of these satellites—the Magellanic Clouds—are considerably larger than the others. The large cloud has about fifteen billion stars, the small cloud about five billion.

The orbits of the Magellanic Clouds are marked by an enormous river of cold hydrogen gas, the Magellanic Stream, in which they swim. At least two other satellites of our galaxy —the Draco and Ursa Major dwarf galaxies—also follow orbits that lie within this river of gas. It is composed of at least six clouds connected by wisps of thinner gas. Each has enough gas to make several tens of millions of stars the mass of the sun. Doubt remains as to the origin of the stream. It may be composed of gas that was drawn from the Milky Way Galaxy by gravitational interactions with the Magellanic Clouds. Or it may be composed of primordial gas that, in a gravitational tug of war between the clouds and the Milky Way, has been permitted to settle into allegiance with neither. Similar cold clouds have been found in the spaces surrounding other galaxies, often in association with their satellites.

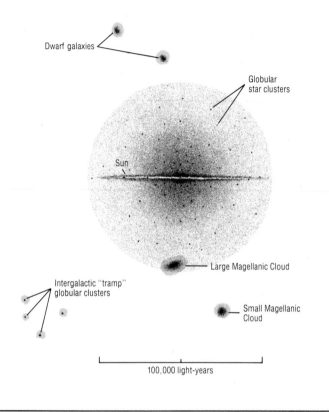

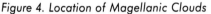

100,000 light-years

Figure 4. Location of Magellanic Clouds
In the outer domain of the Milky Way lie dwarf satellite galaxies and a few distant globular star clusters. The two large satellites known as the Magellanic Clouds lie in the foreground in this perspective, and should be envisioned as hovering a few inches above the page.

36 The large Magellanic Cloud, visible in great detail at its distance of only one hundred and fifty thousand light-years, shows evidence of many features familiar to us in our own galaxy; the ruddy cloud of gas toward one end of the galaxy, the Tarantula Nebula, belongs to the same species as the Orion and Trifid nebulae here in the Milky Way; with a diameter of eight hundred light-years and a mass of perhaps three hundred thousand times that of the sun, it is the largest such nebula known. Were it as close to us as Orion its brightness would rival the moon's.

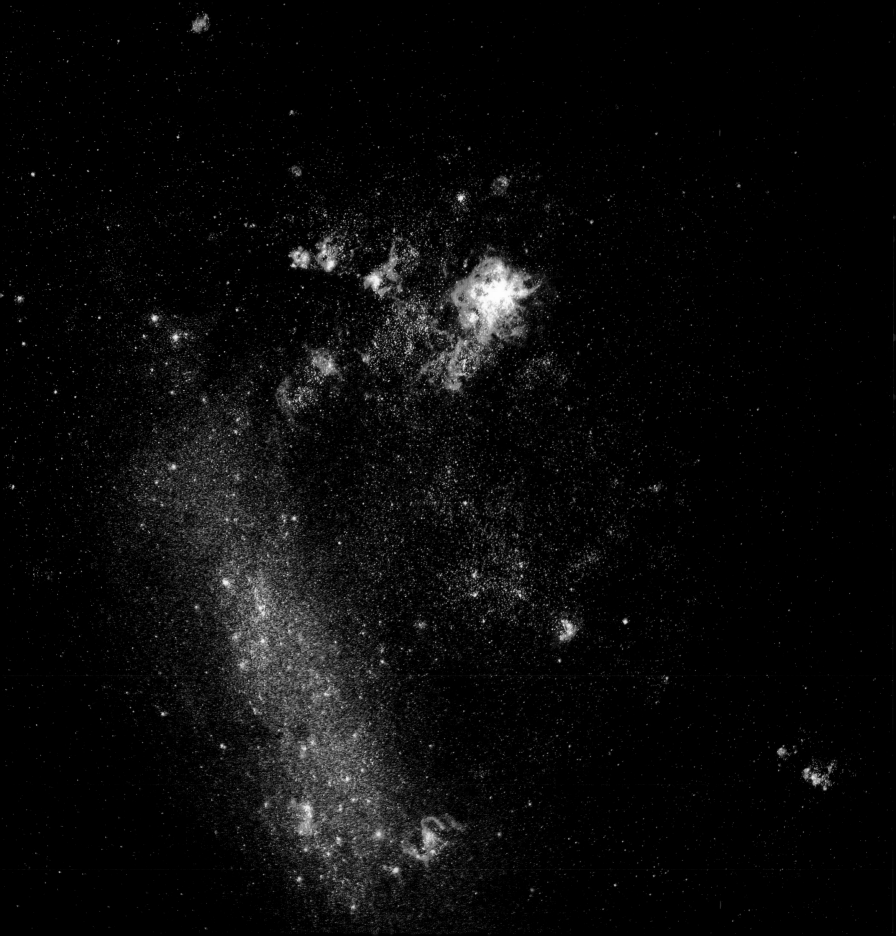

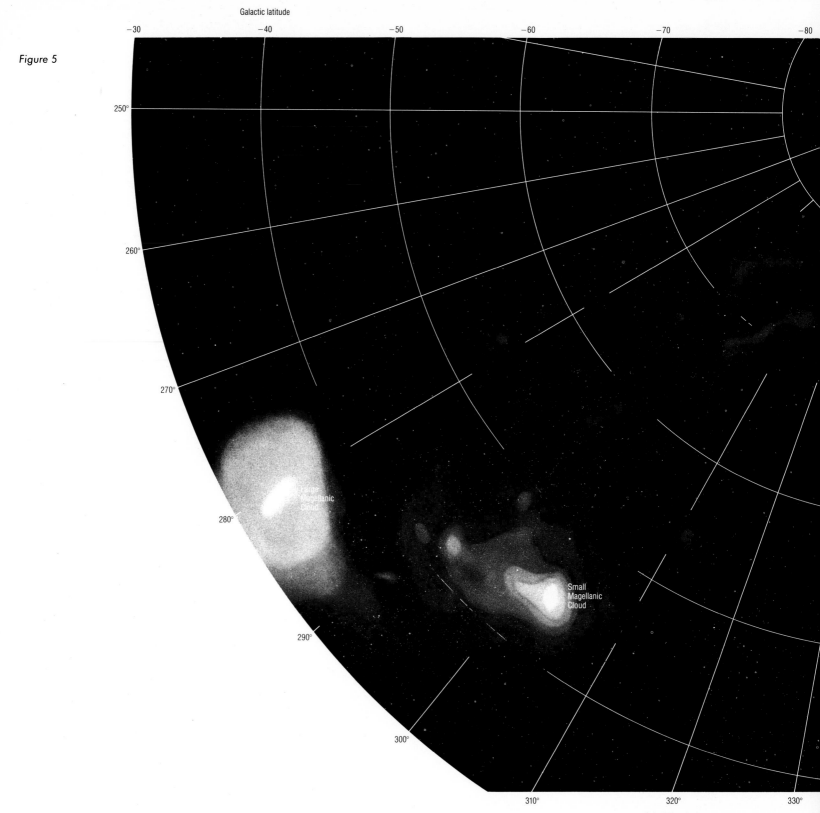

Figure 5

Galactic latitude

−30 −40 −50 −60 −70 −80

250°

260°

270°

280° Large Magellanic Cloud

290° Small Magellanic Cloud

300°

310° 320° 330°

Galactic longitude

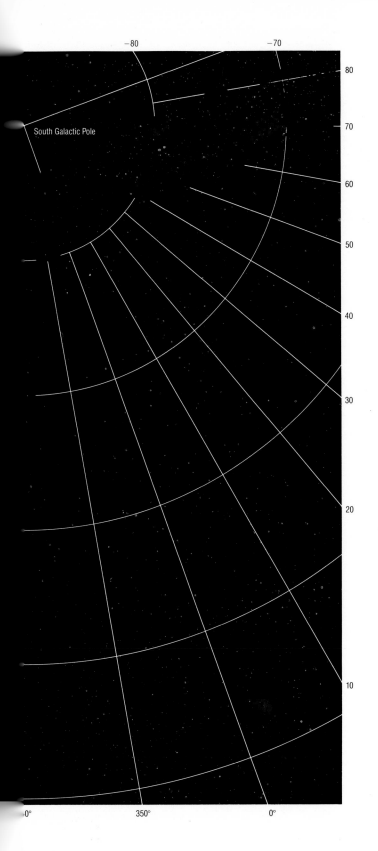

South Galactic Pole

37

Figure 5. The Magellanic Stream
A giant loop of cold hydrogen gas encircling our galaxy, the Magellanic Stream, was so named because it lies along the orbit described by the Magellanic Clouds. The clouds appear to the lower left of this map, which was made by charting radio radiation emitted by the intergalactic gas. The coordinates are those of galactic latitude and longitude.

37 Riding at close quarters to each other, the Large and Small Magellanic Clouds are perhaps destined to merge into a single galaxy.

38

38 This photograph of the Small Magellanic Cloud, a longer exposure than the color photo opposite, sacrifices some of the exquisite detail of the central regions but reveals more of the outlying stars of this small galaxy. The overexposed blob of light near one edge of the frame is the globular star cluster 47 Tucana, about half as far from us as the Small Magellanic Cloud.

39 The Small Magellanic Cloud (right), a satellite galaxy of our Milky Way, is rich in young, blue stars, while our galaxy's other satellite, the Large Magellanic Cloud (page 71), includes more older, red stars.

The Andromeda Galaxy

The Milky Way and its sister the Andromeda Galaxy constitute an example of one of nature's most grandiose creations, a pair of spiral galaxies. Many spirals belong to such pairs. Usually, the pairs are asymmetrical, like a crab's claws, one galaxy larger than the other. Andromeda, the larger of this pair, has about twice the mass of the Milky Way. The two galaxies rotate in complementary directions, one clockwise and the other counterclockwise, so to speak; this characteristic of their relationship, found in many other pairs of spirals as well, lends support to the hypothesis that the two galaxies formed at about the same time, from two adjacent whirlpools of primordial gas, rather than having formed far apart and later blundering into each other's company.

Similarities between the two galaxies are abundant. Each displays the accoutrements of a major spiral—a central region composed chiefly of old stars, an expansive flat disk populated by tens of billions of stars of widely assorted ages and chemical compositions, dust-laden spiral arms rendered incandescent by the light of newly formed stars, and a spherical halo of old dwarf stars embracing the galaxy and highlighted by hundreds of globular clusters. Each galaxy is attended by two prominent satellite galaxies plus many less prominent ones. It even happens that the plane of each galaxy is inclined to the other's line of sight at almost exactly the same angle, so that the Milky Way viewed from Andromeda ought to look quite a bit like Andromeda viewed from the Milky Way.

Here, while we behold a major galaxy at close range, may be an appropriate time to ponder what it is we are seeing in photographs of galaxies. The photograph records light and only light. Here on earth we are accustomed to seeing things by virtue of their reflected light; that is how we discern a smile or a mango or the moon. Photographs of galaxies offer us little such reflected light. Planets of stars in other galaxies, however many there may be, are too small and inconspicuous for us to see them, and light reflected from dust clouds—a phenomenon found occasionally in our galaxy, as in the case of the Pleiades star cluster—is too dim to show up in photographs like these.

What we do see is the light from billions of giant stars and from the bright nebulae in which some of the stars are embedded. This light reaches us from widely scattered quarters of what is, after all, a staggeringly large system of stars. If we consider that the Andromeda spiral is roughly one hundred thousand light-years in diameter, and note that its inclination to our line of sight sets the near edge of the spiral nearly one hundred thousand light-years closer to us than the far edge, then we face the curious conclusion that the light recorded in the photograph from the far edge is one hundred thousand years older than the light from the near edge. We are looking not only at a great deal of space, but at a thousand centuries of time.

The astronomer and philosopher of science Sir Arthur Stanley Eddington used to say that he wasn't sure he had never seen a star, in that he had seen only the light that comes *from* a star. Similarly, we might say that we have never seen a galaxy, in that we have seen no such single *thing* as a galaxy. What we see is light that informs us that a great many stars are there, and that they are arranged in a certain pattern. We call that arrangement a galaxy. But a galaxy is not a thing; it is a collection of things, or a collection of phenomena.

This is not meant to begrudge our view of the Andromeda Galaxy. It is alive with information and with heart-rending beauty. And it will improve as time goes by, for our galaxy and Andromeda, orbiting around a common center of gravity, are drawing closer together. Every second brings us fifty miles closer to Andromeda. In a few billion years, the two galaxies will be only half as far apart as they are today, and Andromeda will loom twice as large in our skies.

---

40 The Andromeda Galaxy (=M31=NGC224) at a distance of 2.2 million light-years is the nearest large spiral to our galaxy. Its colors result from a predominance of older red and yellow stars in the central regions that give way to a predominance of younger blue stars in the realm of the spiral arms. The stars scattered across the frame all belong to our galaxy and lie in the foreground, like raindrops on a window pane. Andromeda's two prominent satellite galaxies, analogous to the Magellanic Clouds that orbit our galaxy, are M32 (=NGC221), projected against an outer spiral arm, and NGC205 on the opposite side. If this were a photograph of the Milky Way taken from Andromeda, our sun would be located on the inner edge of one of the outermost visible spiral arms.

Maffei I

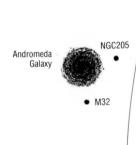

NGC147

NGC185

Andromeda
Galaxy　NGC205

M32

M33

Figure 6. The Local Group
The galaxies of the Local Group are here plotted approximately to scale. The concentric circles mark intervals of one million light-years from the center of the Milky Way galaxy. Notice the pronounced binary nature of the Local Group: Most of its galaxies are clustered around its two dominant members, the Andromeda and Milky Way galaxies. The preponderance of satellites near the Milky Way is probably not a genuine effect, but reflects the fact that these dim satellites are more readily detected in nearby intergalactic space than are those at the distance of the Andromeda spiral. The Group is still far from being adequately mapped, and we may expect that many small Local Group galaxies remain to be discovered.

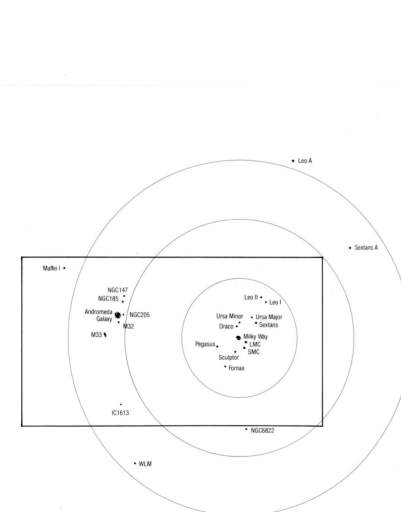

• Leo A

• Sextans A

Maffei I •

NGC147
NGC185
　　Leo II •　• Leo I
Andromeda　• NGC205
Galaxy　• M32　Ursa Minor　• Ursa Major
M33 ᛉ　Draco •　• Sextans
　　　　Pegasus　• Milky Way
　　　　　• LMC
　　　　　• SMC
　　　Sculptor
　　• Fornax

•
IC1613

• NGC6822

• WLM

IC1613

84

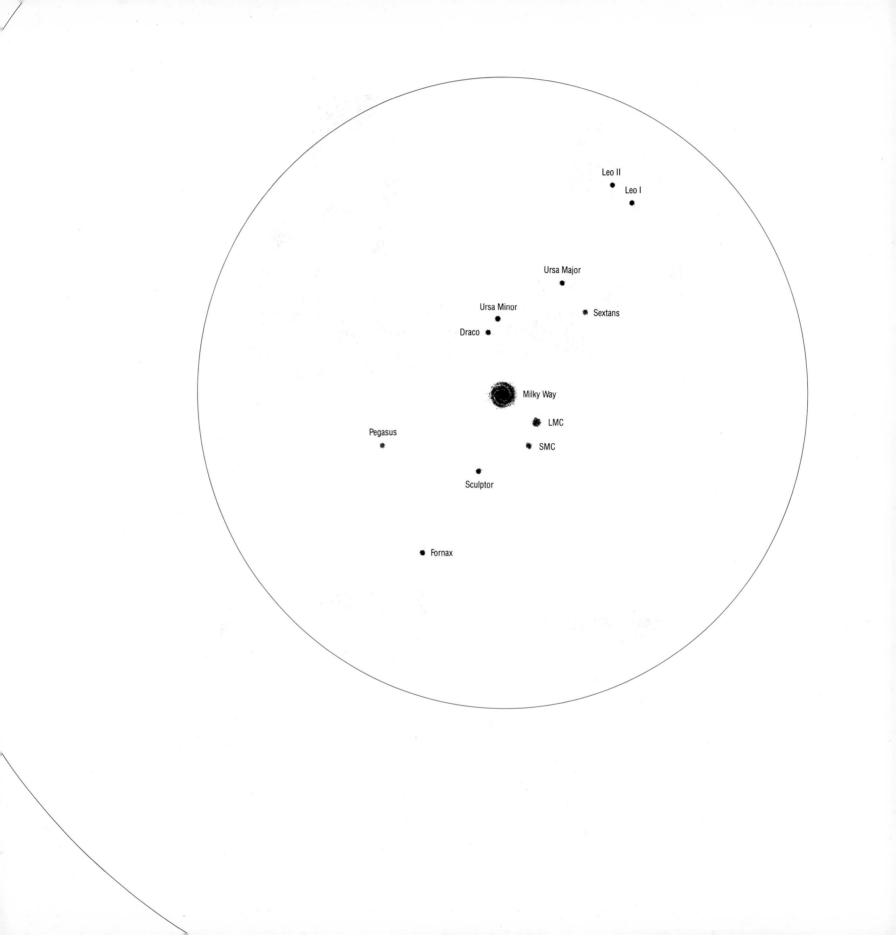

Leo II

Leo I

Ursa Major

Ursa Minor

Sextans

Draco

Milky Way

LMC

Pegasus

SMC

Sculptor

Fornax

The Sculptor Dwarf Galaxy

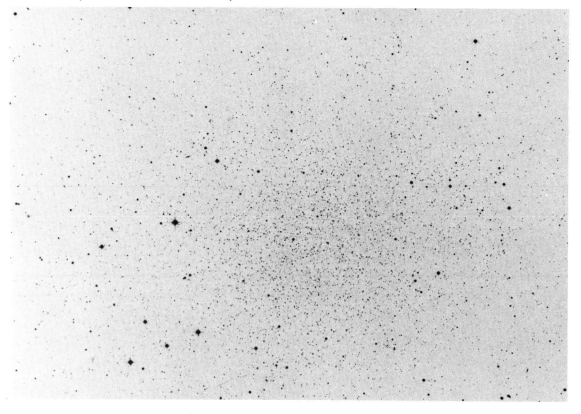

52 The sparse stars of the Sculptor Galaxy, a dwarf only about eight thousand light-years in diameter, are displayed here in a negative print that renders stars black and the sky white.

In the heavens as on earth we find plentiful evidence of what Shakespeare called "the modesty of nature," her predilection for the unassuming, the inconspicuous and the small. On earth we find more plankton than great fishes, more insects than human beings, dirt ubiquitous, gold rare. In our solar system, the most numerous objects are not giant planets like Jupiter and Saturn, or even modest planets like the earth, but the hoards of tiny planetlets called asteroids, few of them so large as a steamer trunk. In our galaxy, the majority of stars are dwarves. And among the galaxies themselves, the majority are not great systems like the Milky Way or Andromeda; they are the dwarf galaxies.

The Local Group contains dozens of dwarf galaxies. They are so paltry that altogether they contribute less than a tenth of the mass of the group as a whole.

The stars of the Sculptor Galaxy (above) add up to the equivalent of only about two million stars the size of the sun. The galaxy is so loosely organized that we can see right through it. It has no visible nucleus, and lacks the interstellar clouds characteristic of the spirals.

This deficiency of interstellar clouds may be due to the relatively low gravitational attraction of a galaxy as meager as Sculptor. Large galaxies are thought to garner interstellar material in two ways, by vacuuming up intergalactic gas as the galaxy moves along, and by the shedding of material by their stars in the form of stellar winds, "planetary" nebulae, novae and supernovae. But neither of these methods works well for a galaxy as small as Sculptor. Its gravitational force is insufficient to scoop up much gas from intergalactic clouds, some of which are more massive than Sculptor itself. And it is at a loss to retain much of the ejecta of its own stars; when, for example, a giant star in Sculptor explodes, the shell of its ejected material simply flies out into intergalactic space.

The Sculptor Galaxy is a satellite of the Milky Way, and orbits it in roughly the same path that is followed by the Magellanic Clouds and is demarked by the river of gas called the Magellanic Stream. There is evidence to suggest that a number of dwarf satellites of the Milky Way lie along a sort of great circle route defined by the Magellanic Stream. if so, our galaxy is being orbited by a chain of little galaxies.

III/The Form and Variety of Galaxies

*For those who are awake
the cosmos is one.*

—HERACLITUS

A Journey through Intergalactic Space

We accelerate out of the Local Group. As our speed edges ever closer to that of light, time on board passes ever more slowly by comparison to that of the universe at large. Seen through our eyes, the cosmos up ahead conducts its affairs with crazy haste. Planets whirl in their orbits. Stars are formed and die between breakfast and supper. We speed on.

We have entered upon the deep spaces that intervene between groups of galaxies. The familiar spirals of Andromeda and the Milky Way have shrunk until smaller than a fingernail at the distance of an outstretched hand. Free from nearby distractions, we are left for once in the sole and equitable company of everything—all the galaxies—floating in space in all directions. We while away our time by examining them through the ship's telescope.

What we see reminds us of home.

Back on earth, we recall, all the things of nature shared a deep kinship. Objects as dissimilar as snowflakes and stones proved to be made up of combinations of atoms drawn from a common pool of elements. Living things as diverse as a boll weevil and a human being were found not only to be made from the same common stock of atoms, but to have been built to the design of a single sort of molecule, that of DNA. All the creations of our planet could be understood as having been formed within the parameters of a few fundamental principles of physics. Yet for all their kinship, no two things could be found exactly alike—no two identical snowflakes or stones, boll weevils or people. Nature's way seemed to be to try everything without ever doing the same thing twice.

Now we find this way at work among the galaxies as well. All function within the purview of basic physical principles. Each galaxy, for example, must move through space along the trajectory dictated by its gravitational interaction with its neighbors and with the matter of the universe at large; no galaxy can pick up its skirts and scamper away in violation of those laws. And the material kinship of galaxies runs deep. All the stuff of all of them is made, so far as we can see, from various mixtures of the same sorts of atoms that we came to know back on earth.

Indeed, order and regularity are sufficiently manifest in the appearance of galaxies that we can sort them into categories.

About half the galaxies we see are spiral in form, like the

Milky Way. Some among our crew take chauvinistic pride in learning that their home galaxy is of the sort most widespread in the cosmos. Others more dispassionate point out that since most stars are to be found in spiral galaxies, the odds are that any given species evolving on a given planet circling a star will find itself in a spiral, as did we.

About one quarter of the prominent galaxies are ellipticals. Here the stars are arranged not in the flattened disk characteristic of spirals, but within a more nearly spherical volume of space. At first the ellipticals may look rather bland to our spiral-accustomed eyes, but as we study them further we come to appreciate their symmetry of form, their purity of content (ellipticals contain little interstellar gas and are made chiefly of stars and space), and the magnificence of their most exemplary representatives, which number among the largest galaxies in the universe.

Scattered among the many other galaxies we find the SO, or lenticular, galaxies, much like spirals in form but lacking spiral arms. They combine some of the qualities of both spirals and ellipticals.

A few percent of the major galaxies are irregular. Their virtues are those of individuality, even of eccentricity. In their splayed and contorted forms they offer us endlessly varied perspectives on the star fields and nebulae they contain, like translucent sea creatures, whose interiors and exteriors may at once be seen.

Dwarf galaxies abound, most of them ellipticals and irregulars. Frequently we find them ranged around larger galaxies. If they seem negligible by comparison, we need only consider that even a dwarf contains millions of stars.

If no two galaxies are identical, no two stars or planets identical, then how can we imagine the variety manifest in the universe on a planetary level? Is there to be found across the whole sweep of creation a single insect, flower, raindrop or mud puddle that somewhere has a twin? And where thoughtful life has arisen, to what degree do its imaginings converge with that of other intelligences, in consequence of nature's predilection for order and form, and to what degree do they diverge, in consequence of nature's predilection for variety?

What is the cosmology of imagination, we wonder as our imaginary ship wanders on.

III/THE FORM AND VARIETY OF GALAXIES

Normal Galaxies

SPIRALS

Most large galaxies are spirals. Their general anatomy can be described in terms of three components: a central region, elliptical in shape and centered upon the nucleus; a broad, flat disk shared by stars and interstellar clouds; and a spherical corona, or halo, composed primarily of old dwarf stars and globular clusters and embracing the galaxy as a whole. If we arrange spiral galaxies formally in terms of the size of the central bulge relative to the disk of each, we find that they can be placed along a continuum. A classification scheme based upon such an arrangement and widely in use among astronomers is illustrated on page 91. In this system, galaxies with large central regions are classified Sa, intermediates Sb, while spirals with small central regions relative to their disk are classified Sc and Sd. This one-dimensional scheme falls far short of accounting for all the important parameters of form in spiral galaxies, and debate continues about its details, but the fact that even a partially coherent system of classification is possible encourages the expectation that we may one day come to fully understand how matter came to be deployed across the cosmic theater in the form of galaxies.

Many elements of continuity have been discerned within the classification of spiral galaxies. The central regions, the bulges, of spirals are populated predominantly by old stars and are rare in interstellar material; in some ways they resemble elliptical galaxies like those pictured on pages 118 and 119. The disks of spirals are relatively rich in interstellar material; in many cases, new stars are being formed in the disk today, and as a result of this ongoing production of stars the stellar population of the disk is far more heterogeneous in terms of the ages of its stars than is the central region. Generally speaking, Sc galaxies with their relatively larger disks are more active in creating new stars, while Sa galaxies create stars more fitfully and less abundantly. The Milky Way Galaxy, an Sb or perhaps Sc located roughly midway between the extremes, we would expect to be populated by a heterogeneous mixture of old, middle-aged and young stars. And this is just what we do find here. Our sun is one of the middle-aged stars.

Some of the interstellar material in the disk of a spiral galaxy consists of dust that has been processed through the cores of stars that subsequently exploded, divesting themselves of material that included many of the heavier elements we call metals. The halo, composed primarily of stars that formed before there had been time for many of the heavy elements to be created, is generally metal-poor. The disk with its many younger stars is typically one hundred times richer in metals. A gradient in metal abundances may be detected across the disk itself: There are fewer metals in the more thinly populated outer disk—more metals in the interstellar inner spiral arms where the stellar population is dense and where more stars have been born and died. It should be remembered that these are generalities, like stating that St. Petersburg, Florida, has an elderly population (though not everyone in St. Petersburg is elderly), or that the Atlantic Ocean is salty (though the salinity of the ocean varies considerably from one spot to another).

The dynamics of spiral galaxies are elegant and subtle. Nothing illustrates this better than the spiral arms themselves.

We can readily see that spiral arms cannot be objects, like vines or tree branches. A spiral galaxy does not rotate all of a piece, like a phonograph record, but rather rotates differentially: Stars near the central regions orbit the galaxy much more rapidly than do stars in the outer reaches. In a typical spiral, disk stars near the central bulge complete an orbit in about twenty million years, while those in the outer precincts of the galaxy take some two hundred million years, ten times longer to complete one "galactic year." Much the same is true of the interstellar material of the disk; its rotation is likewise differential. If the spiral arms were all of a piece, differential rotation would quickly either fragment them or wind them tightly around the galactic center. The situation resembles that of runners confined to lanes on a track. Although they start abreast as they circle the track those on the inner lanes draw ahead. Soon the line of runners must either fragment as the outer runners fall behind, or, if the runners are to stay together, the outer ones must draw in toward the center until all are running on the inner track. Spiral galaxies have maintained their arms for billions

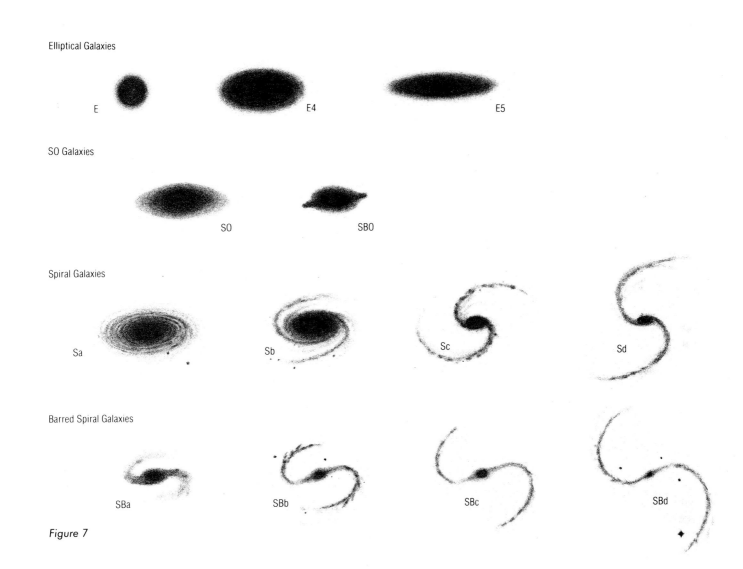

Elliptical Galaxies

E   E4   E5

SO Galaxies

SO   SBO

Spiral Galaxies

Sa   Sb   Sc   Sd

Barred Spiral Galaxies

SBa   SBb   SBc   SBd

Figure 7

of years without either fragmenting them or wrapping them up, so we must abandon the view that they are physical entities and search for a mechanism that will explain them as a phenomenon.

The theories that have done this most successfully are those that view the arms as density waves propagated through the interstellar medium, like ripples in a pond. The density waves are set up by resonances in the gravitational interaction of the galaxy's billions of stars in their orbits.

We are able to see the spiral arms by virtue of the fact that when a wave encounters an interstellar cloud it often raises the density of portions of the cloud sufficiently for stars to form from the collapsing cloud. Their light, and that of the surrounding clouds that they illuminate, traces out the contours of where the density waves recently passed, and it is this luminous phenomenon that we see as the arms.

Figure 7. Galaxy Type Illustration
The formal continuum of galaxies is here illustrated schematically, with the spiral scheme extended to include galaxies with very small relative central regions classified Sd. Elliptical galaxies are classified E0 through E5 in accordance with their degree of flattening. Notice that perspective effects can greatly influence the classification of ellipticals, in that even a cigar-shaped E5 will look like an E0 if we happen to view it end-on. The SO galaxies retain a category of their own, pending a better understanding of how they ought to be placed in the scheme of things.

55

56

59

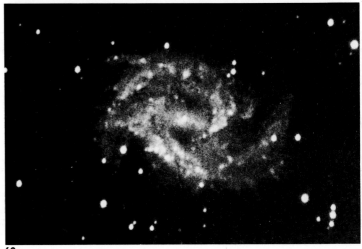

60

in this story. Galaxies are so large and their phenomena transpire over periods of time so lengthy by human standards that in the early years following their discovery we may have been guilty of thinking of them as static entities, like stuffed birds. Now we are coming to realize that they more nearly resemble birds in flight.

While susceptible to formal categorization, spiral galaxies display wonderful idiosyncracies of form, and it is this variety that permits even the casual viewer to assign to each spiral a distinct personality. The Sb galaxy NGC2841 (page 96) at first glance appears to display the classic spiral pattern of two dominant arms winding themselves several times around the system. But looking more closely at the photograph, we

find that this structure is composed not of continuous arms but of a series of filaments, each composed of scores of bright nebulae and blue stars.

Even the most pronounced spiral structures prove upon careful inspection to be complex and individualistic. So seemingly straightforward a spiral as M101 proves, if we attempt to trace out its spiral arms, to be as fractured and jumbled as the geology of a young mountain range. At several points the spiral arms cross over each other, a situation not easy to account for in terms of our tentative theories of spiral arm formation.

While the amount of interstellar gas and dust in elliptical galaxies typically amounts to less than one one-thousandth

57

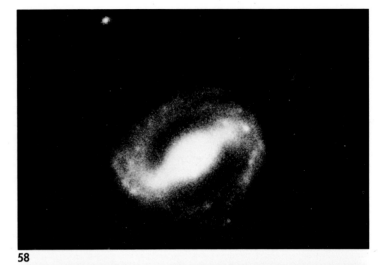

58

61

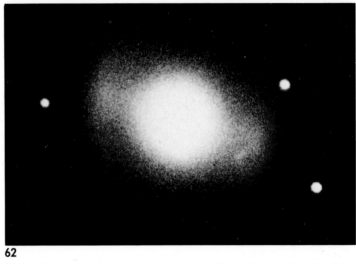

62

of their mass, the interstellar material in spirals may account for several percent of their mass, with the fraction increasing as we move down the spiral sequence from Sa to Sd. Irregulars include some of the murkiest galaxies, with up to fifty percent of their mass in the form of interstellar clouds.

A galaxy can acquire interstellar gas and dust from three known sources. First there is the primordial hydrogen and helium gas, atoms as old as the universe that were incorporated into the galaxy when it formed and have as yet escaped being incorporated into stars. Second is gas and dust that once were part of stars and subsequently were ventilated back into space by means of stellar winds, the ejection of material in novae of "planetary" nebulae, or the explosion of stars as supernovae. This processed material

55–62 Spiral galaxies may be divided formally into barred and nonbarred spirals, and organized within each category on the basis of the size of their central bulge relative to the arms. In this sampling the Sa galaxy NGC2811 (**55**) is characterized by a prominent central region, while the Sb galaxy NGC2841 (**56**) has relatively more prominent spiral arms, and the Sc galaxy M74 (**57**) has a relatively small central region. Barred spirals, as the name implies, feature a conspicuous barlike congregation of stars and interstellar material projecting from the central region; the SBa galaxy NGC175 (**58**) displays a large bar and central region; the prominence of these features diminishes through the SBb system NGC1300 (**59**) to the SBc NGC2525 (**60**). The SO galaxies resemble other spirals but lack any evidence of spiral arms; NGC1201 (**61**) is a typical SO, while NGC2859 (**62**) is classified as a barred SO, designated SBO.

63

includes heavy atoms built in the interiors of stars—the dust component of interstellar clouds. The third way a galaxy can acquire interstellar material is by sucking it in from intergalactic space; this infalling process is believed to play an important role in galactic ecology, but much remains to be learned about it.

Dust-bearing interstellar clouds become conspicuous when silhouetted against background stars. The rift of dust that stretches across the central regions of M64 (page 98), as striking as a furrow plowed across a bare field, makes a good example. The cloud is so thin that if we were to scoop up a jar of it and analyze it in a laboratory we would have

64

difficulty telling it from a perfect vacuum, but it is so large that its lonely vagrant atoms add up to an enormous amount of material, enough to topsoil the gardens of billions of planets. Walking barefoot on earthly soil, we might contemplate that every atom of the dirt beneath our feet once drifted in space in clouds like these.

63 The spiral galaxy NGC2841 is classified Sb (above).

64 A spiral arm structure as intricate, if not so fragmented, as that of NGC2841 characterizes this large spiral galaxy, NGC2613 (above).

65

66

65, 66 M64 (= NGC4826) **(65)** and NGC3623 **(66)** are Sb galaxies with conspicuously heavy clouds of dust and gas marking their disks.

67 The Sombrero Galaxy, M104 (= NGC4594), an Sa galaxy, ought to have relatively little interstellar gas and dust, yet the dark silhouette of its plane indicates that it is rich in interstellar material.

68

The interstellar clouds of spiral galaxies tend to take on a disheveled appearance. The gravitational and magnetic fields of the galaxy tug at them, density waves compress them, starlight and stellar winds waft through them, and the clouds collide with one another.

Upwellings of interstellar clouds resulting from this interplay can be seen most evidently in galaxies that present themselves to us edge-on, as does NGC4565 (above). Two remarkable geysers, rising up from the plane of the galaxy, are silhouetted against the starlight of the central bulge. Off to the left a lofty pair of arches may be seen; these appear to consist of material that has been squirted up out of the disk and now, responding to the gravitational pull of the disk, is returning to it, like a spent skyrocket returning to earth. Celestial festoons like these may achieve altitudes of hundreds of light-years.

---

68 NGC4565 (above), tipped only four degrees from a perfect edge-on perspective, exhibits the components of a normal spiral galaxy—the elliptical central region and flat, dust and gas-laden disk.

BARRED SPIRALS

About one-third of all spiral galaxies are "barred" in form to a pronounced degree. By "bar" is meant a spindle-shaped grouping of stars and interstellar material that extends outward to either side of the central bulge and from which the spiral arms stem. Its length is typically a few tens of thousands of light-years. As the central bulges of many ordinary spirals seem to be somewhat oval or elongated in shape, it is possible that most spirals contain at least a vestigial bar. In the most widely accepted system of galaxy classification, the barred spirals are designated SB and are arranged according to the size of their central bulge along a continuum SBa to SBd, extending to irregular barred spirals like the Magellanic Clouds, designated SBirr.

Why some galaxies have bars is, like so many elementary questions about galaxies, still unanswered. One possibility is that the bar represents a way for the stars of a galaxy reared in disordered surroundings to settle into relatively stable orbits, stabilizing the galaxy in something like the way a tightrope walker regains his balance by extending his arms.

The barred spiral M83, one of the most dynamic-looking galaxies in the sky, looks almost as if it were tumbling, like a child's top kicked across the floor. In the absence of a photograph taken from another perspective by an observer in another quarter of the universe, we must try to decipher its three-dimensional shape as best we can from the vantage point granted us. So studied, M83 appears to have been warped, portions of its spiral arms bent well out of their original plane. Perhaps if we could see it edge-on it would resemble the "can opener" profile of NGC2146 (page 125).

Radio maps like the one on page 105 show M83 to be surrounded by enormous puddles of cold hydrogen gas. The form of these clouds suggests that of vestigial spiral arms lagging behind the arms of the galaxy within. This is just what we would expect to find had the galaxy indeed been warped relative to its surrounding envelope of gas. The galaxy would have gone on rotating, while the gas clouds, freed from some of the gravitational dominion that had been exerted upon them by mass concentrated along the plane of the galaxy, would have lagged behind. The gas cloud may

69

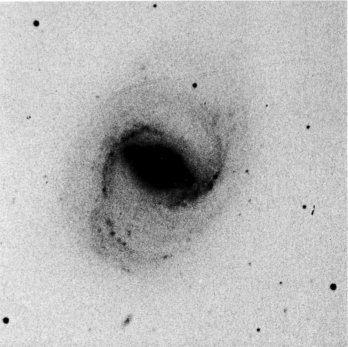

70

69, 70 The galaxies NGC3992 (**69**) and NGC4541 (**70**) are barred spirals; the "bar" is usually shaped something like a spindle, as the negative print of NGC4541 helps make clear.

101

71

72

73

---

71 The galaxy NGC1360 (left) is also a barred spiral.

72, 73 The barlike structures developed in many galaxies like these, NGC4650 (**72**) and NGC4548 (**73**), are thought to be dynamically stable forms into which stars may organize in a galaxy that has suffered from initial perturbations.

describe the old plane of M83, out of which the galaxy has since been wrenched and now sits like a barge listing in the sea.

What created the list? The only other galaxy in the vicinity of M83 is a small elliptical, NGC5253. It has only about one-tenth the mass of M83 and so ought not to have been able to precipitate so marked a disturbance, unless the two systems virtually collided.

LENTICULAR GALAXIES

Difficult to categorize are the SO, or lenticular (meaning lens-shaped), galaxies. Though shaped like spirals in that they have a central bulge and a large thin disk full of stars, unlike the spirals they have little interstellar dust and gas and no spiral arms. They might be described as elliptical galaxies cast in a spiral mold, and are sometimes said to represent an intermediate step between ellipticals and spirals. It may be more to the point to say that SOs look like nothing so much as spirals that have been robbed of their complement of interstellar gas and dust.

If so, what robbed them? According to one hypothesis some SOs are created when a normal spiral blunders into another galaxy or into a massive intergalactic gas cloud; such an encounter ought to leave the spiral in possession of its stars but swept clean of its interstellar material. Proponents of this explanation point out that SO galaxies are found most frequently in the inner regions of clusters of galaxies, where such collisions ought to occur most frequently. Opponents of this view point out that SO galaxies have been found drifting alone in free space, well away from any clusters; if intergalactic collisions were responsible for stripping their interstellar material, with what did they collide?

ELLIPTICAL GALAXIES

Elliptical galaxies present a simpler appearance than that of the spirals. In place of the mutli-component nature of a spiral galaxy—nucleus, central bulge, disk and corona—ellipticals are simply a case of billions of stars gathered in a roughly spheroidal volume. Most have not even a nucleus. The pristine clarity of the space between their stars is sullied by only the scarcest traces of interstellar material. A halo of globular clusters is customarily an elliptical galaxy's sole concession to ornament.

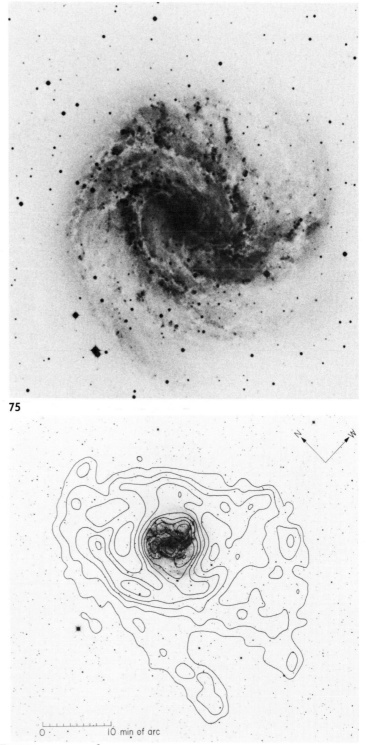

75

76

74 Alive with motion, M83 (= NGC5236) typifies the dynamic processes apparent in spiral galaxies.

75 The fanlike structure of interstellar clouds between the arms of M83 may be seen to advantage in this negative print.

76 Vast puddles of intergalactic hydrogen gas surround M83, here traced with a radio telescope at the twenty-one-centimeter wavelength.

105

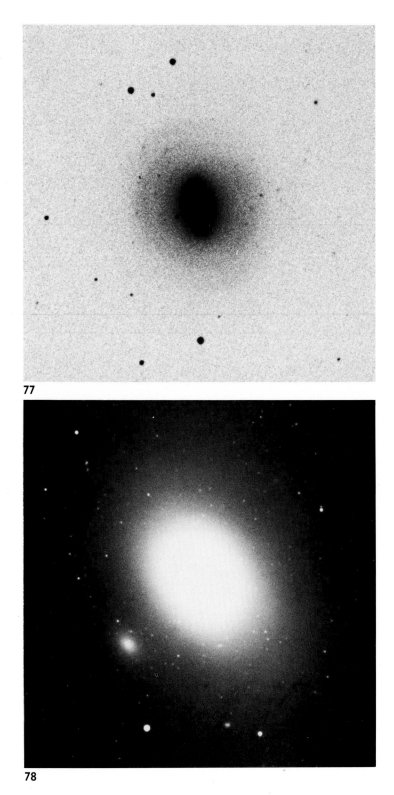

77

78

No perfectly spherical elliptical galaxy has ever been found, but some are nearly spherical. These are classified EO. Others, more flattened, take on shapes resembling squashed pincushions. These are categorized E1 through E5, according to their degree of flatness. The effects of perspective confuse the business of determining just how flattened an elliptical galaxy is, since even a radically flattened elliptical system will appear circular if we happen to be viewing it from the direction of one of its poles. As the astronomer Sir Fred Hoyle points out, an elliptical galaxy is always at least as flattened as it appears to be.

Unlike the stars of a spiral galaxy, which generally follow orbits that lie along the plane of the disk, like runners rounding a track, the stars of elliptical galaxies pursue orbits that are inclined at a great diversity of angles. Their orbits resemble the flights of hunting sea birds, some diving and then swooping upward while others circle variously amid them. A few elliptical galaxies are thought to be rotating as a whole; others display no evidence of rotation.

Spiral galaxies all lie within a relatively constrained range of mass, most having the equivalent of between ten billion and a few hundred billion stars like the sun, but ellipticals are far more varied in terms of mass. Dwarf ellipticals with only a few million stars and a diameter of but a few thousand light-years are common, while at the other end of the scale supergiant ellipticals have been found with populations estimated at ten thousand billion stars.

Unphotogenic, the ellipticals tend to be underrepresented in books of photographs like this one, though they constitute perhaps twenty percent of the prominent galaxies in the known universe. If we lived in an elliptical galaxy, however, we might cherish its Apollonian simplicity of form, so unlike the tangle of the spirals, and be grateful for the lack of interstellar clouds like those that block our view of much of the sky here within the Milky Way spiral. And we might take pride in a lengthy celestial history. Although ellipticals and spirals are estimated to be about the same age—roughly ten to fifteen billion years old—the ellipticals seem to have turned most of their raw material into stars quite early and thereafter gotten out of the star-making business. Today, the stars in

77, 78 The galaxy M84 (= NGC4374) (**78**), classified as an SO by some astronomers, as an elliptical by others, demonstrates the difficulty of galactic categorization. NGC4477 (**77**), on the other hand, is unambiguously an SO. Even when viewed in the detail of a negative print, its disk displays only the faintest traces of spiral arms.

ellipticals are predominantly old. They glow the dull orange of antique lamps, and that in a sense is what they are. Let us accord them the respect owed to elders. They were ablaze with the light of young stars, their planets were basking in that light, their stories were unfolding, when the earth and the sun were less than a whirlpool in a cloud.

IRREGULAR GALAXIES

Irregular galaxies introduce a touch of disorder into a cosmos otherwise dominated by the ethereal beauty of the spirals and the bald symmetry of the ellipticals. It is thought that they may be whipped into their disordered states in any of several ways. Many are satellites of larger galaxies; here it is clear that each might regain a classical form if it could spend some time away from the disruptive gravitational interference of the dominant galaxy. The Large Magellanic Cloud is an example; free from the Milky Way, it might be expected to reorganize itself into a more symmetrical form. Others may be transformed into irregulars by near collisions with passing galaxies. The disturbed appearance of NGC5195, recently disarrayed in an encounter with M51 (page 131), attests to this possibility. There may be other ways to create an irregular galaxy, but most irregulars seem to be little galaxies that have been bullied by bigger ones.

OUR PERSPECTIVE ON GALAXIES

The stars we see scattered across extragalactic photographs lie in the foreground and belong to our galaxy, not to the remote galaxies being photographed. We peer out at the universe through these scrims of stars, something as our remote ancestors viewed the world each morning from tree-branch perches through a foreground clustered with leaves.

The average density of the foreground stars varies considerably depending upon the part of the sky that intervenes when we train a telescope upon a particular galaxy. Foreground star fields generally are least dense where we are looking out at angles roughly perpendicular to the plane of our galaxy. The nearer our line of sight comes to the

79 M49 (= NGC4472) is an elliptical galaxy classified E4.

80 The Carina dwarf galaxy, difficult to distinguish from the many foreground stars in the Milky Way that lie along our line of view, is a dwarf elliptical.

81

82

83

84

81 NGC3077, an irregular galaxy, is gravitationally enslaved to the large spiral M81 (see page 136).

82 The Sextans irregular galaxy is an outpost member of the Local Group.

83, 84 The Sc galaxy NGC5364 (**83**) lies well away from the plane of the Milky Way, so that we see it with few foreground stars intervening, while the similar Sc galaxy NGC6744 (**84**) lies along a line of sight that passes closer to the Milky Way, where more foreground stars intervene.

85

galactic plane, the more foreground stars there are likely to be. When we try to look out along the plane of the galaxy —that is, through the Milky Way itself—we encounter so many stars that they all but clog the field of view, while massive interstellar clouds conspire to block our line of sight and obscure whole quarters of the universe from inspection at optical wavelengths.

NGC6744 (page 109) lies along a line of sight only twenty-six degrees from the plane of our galaxy, so we glimpse it through a robust thicket of stars. In contrast, NGC5364 (page 108) lies some sixty-three degrees from the plane of the Milky Way, and as a result relatively few foreground stars interrupt our view of it.

Galaxies are sometimes said to be "in" a constellation. Contemporary star charts divide the entire sky into constellations, their boundaries drawn so as to enclose the configurations of bright stars that our forebears named after gods, animals and other figures that caught their fancy or aided their memory. NGC6744 is in Pavo, the Peacock, a southern constellation catalogued in 1603 by the lawyer and

86

87

astrologer Johann Bayer. NGC5364 lies within the boundaries of Canes Venatici, the Hunting Dogs, an asterism once incorporated into the neighboring constellation Ursa Major but promoted to independent status by the seventeenth-century astronomer Johannes Hevelius of Danzig. Inasmuch as the stars of these constellations are close at hand compared to the enormous distances of the galaxies, galaxies may be said to be "in" the constellations only in the sense that the moon, seen through a window, may be said to be "in" the window.

85, 86, 87 The appearance of a spiral galaxy is influenced markedly by the angle at which we happen to view it. A spiral oriented face-on to us as is M74 (= NGC628) (**85**) affords us a view of the full articulation of its spiral arms. Nearly edge-on spirals such as NGC2683 (**86**) and NGC5907 (**87**) deny us much of a view of the structure of the arm, but recompense us by displaying something of the magnificent complexity of structure in the interstellar gas and dust lanes arrayed along the plane of the disk. M74 is classified Sc, NGC2683 as Sb, NGC5907 as Sc.

Supernovae, exploding stars of the highest order of violence, achieve such a brilliance as to be capable of capturing the attention of astronomers—such as there may be—throughout thousands of galaxies. One or two supernovae occur each century in a typical large spiral galaxy.

The light from each of these cataclysms rushes outward into intergalactic space, bringing news of the event to other galaxies with the speed of light. Observers in a galaxy five million light-years away will see the explosion after five million years have elapsed; those in a galaxy ten million light-years away will see it after ten million years. Therefore, the date that a given observer assigns to a given supernova depends upon his distance from it. In this, supernovae serve to remind us that intergalactic distances must be thought of in terms of time as well as space.

The supernovae marked by arrows in the two photographs

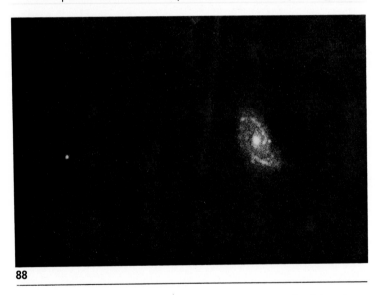

88

88 Extreme depth of field characterizes this photograph of the planet Pluto passing in the foreground of the galaxy NGC5248. When the photograph was taken, Pluto was a little under three billion miles, or 4.18 light-hours, distant. NGC5248 is some seventy million light-years away. The ratio between these distances is approximately the ratio between the diameter of the period printed at the end of this sentence and the distance from New York City to Sydney, Australia. Several dozen star images present on the original photograph have been removed from this print to facilitate identification of Pluto, a planet so small and remote that it appears virtually starlike in even the largest telescopes.

(page 113) were photographed here on earth in the same year, 1961. The galaxies in which they occurred, however, are different distances away from us. The spiral seen more nearly edge-on, NGC4096, is about forty million light-years away; light from its supernova had been traveling for forty million years when it reached earth. The more open spiral, NGC4303, is nearly one hundred million light-years away; the light from its supernova had been traveling for one hundred million years when it reached us. So we may say that in terms of a "cosmic" or universal time frame, the NGC4303 supernova must have occurred first, since its light has been moving through space much longer.

But this assertion could be disputed by astronomers in the galaxies involved. Imagine that there are astronomers in Galaxy A. Impressed by the brilliance of their local supernova, they base their calendars upon it, dating it as Year Zero. The light from this momentous supernova sets out on its trek across intergalactic space. Suppose that seventy million light-years separate Galaxy A from Galaxy B; therefore seventy million years must elapse before astronomers in Galaxy B can observe the Galaxy A supernova.

Before light from the supernova in Galaxy A reaches them, the astronomers of Galaxy B have seen a supernova in their own galaxy and similarly have chosen to date it as Year Zero. Ten million years later, light from the Galaxy A supernova reaches them. They record its date as Year Ten Million. The supernova log of Galaxy B therefore reads:

Galaxy B Supernova Log:
Galaxy B supernova, Year Zero
Galaxy A supernova, Year Ten Million

Now the light from the Galaxy B supernova is on its way to Galaxy A. By the time it arrives at Galaxy A, one hundred thirty million years have elapsed there—the sixty million years that had elapsed in "cosmic" time before the Galaxy B supernova occurred, plus the seventy million years required for the light from Galaxy B to reach Galaxy A. The Galaxy A log reads:

Galaxy A Supernova Log:
Galaxy A supernova, Year Zero
Galaxy B supernova, Year One Hundred Thirty Million

This is much the state of affairs with regard to the two supernovae pictured (page 113). Galaxy A is NGC4303, Galaxy B, NGC4096. Astronomers in both galaxies can claim with roughly equal justice that their local supernova was the earlier of the two.

89

90

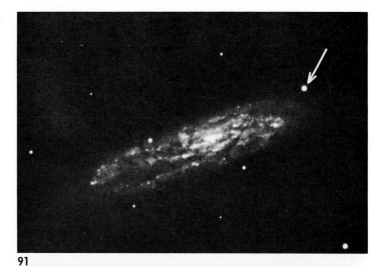

91

92

Here in the Milky Way, light from the two supernovae arrived at about the same time, and so we assign the same date to both—1961. As it happens, we have a local supernova of our own that remains fresh in cultural memory, the one that produced the Crab Nebula (see page 59), and we date it as having occurred earlier than either of these distant supernovae—in 1054 as opposed to 1961. But light from the Crab Nebula supernova has not yet reached either of the two external galaxies, NGC4303 or NGC4096, and will not reach them for millions of years to come. When it does it may be expected to show up on their logs as by far the most recent of the three.

In short, determining when things occurred in the cosmos is very much a matter of where they occurred relative to their observer. Changing our frame of reference will often reorder the temporal sequence we assign to events.

89, 90 The barred spiral NGC4725 is seen during and following a supernova. During its brief prominence, the exploding star shone nearly as brightly as all the billions of stars in the central regions of the galaxy.

91, 92 The exploding star marked by the arrow in NGC4096 (**91**) blew up sixty million years later than the one in NGC4303 (**92**) according to one scheme of reckoning time, but the light from the two explosions reached earth during the same year owing to the fact that the latter galaxy is sixty million light-years farther away than the former. Observers in each galaxy would date their local stellar explosion as the earliest.

Violent and Peculiar Galaxies

All galaxies emit energy—that is why we can see them—but some emit much more energy than do others. These are the "violent" galaxies, sometimes referred to as "active" or even "exploding." The energy output of an active galaxy can be prodigious, and may manifest itself not only in the wavelengths of the electromagnetic spectrum we call visible light, but elsewhere along the spectrum as well—at the longer wavelengths of radio and of infrared light and at the shorter wavelengths of ultraviolet light, X-rays and gamma rays. In many cases, the locus of the galactic energy wellspring is that mysterious region, the nucleus.

The nuclei of active galaxies often show signs of hectic internal motion, their stars and interstellar clouds whirling about. They may be shrouded in thick dust, calling to mind a stop-action photograph of an explosion. And often these nuclei do appear to be exploding, in the sense that they are hurling matter out into intergalactic space.

The output of some of the more profligate violent galaxies would seem to require converting the equivalent of millions of stars like the sun into pure energy, in a cosmic crucible of unknown design. Where does this energy come from?

Adding to the mystery, the brightness of the nuclei of some violent galaxies varies. Some flicker over intervals as short as a few days or weeks; tremendous amounts of energy are being radiated here from something extremely small by galactic standards, perhaps only a light-week or so in diameter.

Various theoretical models have been put forth to explain what is happening in the nuclei of violent galaxies. It has been argued that their energy comes from collisions of stars, from a chain reaction of supernovae, or from a black hole that occupies the center of the galaxy, where it eats stars and interstellar clouds. But problems have arisen in trying to match any of these models to all the violent galaxies. For each violent galaxy that appears to fit a given model, another can be found that does not. One theory may explain how the nucleus produces so much energy, but fail to account for how it varies in brightness. Another may account for the variability, but not for the ejection of material from the nucleus into space. The galaxies in their wildly various behavior have something to tell us, if only we can learn how to better understand their language.

One promising clue lies in the very enormity of the violence. The fireworks of a violent galaxy are so spectacular that the galaxy cannot be expected to have sustained them indefinitely. To have done so, it would have had to convert all its matter into energy and have disappeared long ago.

This insight suggests that violent activity in galactic nuclei is a passing phenomenon, that it may be not a permanent attribute of a galaxy, like a person's having brown eyes, but a passing condition, like having the measles. To entertain this hypothesis requires that we consider the antiquity of the universe and the brevity of our tenure in it.

Imagine that you are looking out across a meadow full of

93

93 A longer exposure in black and white makes visible something of the extensive stellar corona of Centaurus A, a massive galaxy with perhaps three hundred billion stars.

94 Centaurus A, a powerful source of energy in optical, radio and other wavelengths, looks like an elliptical galaxy wrapped in the remnants of a spiral galaxy, and perhaps that is just what it is.

fireflies on a summer evening. Charmed by the sight of the fireflies, you make a snapshot of the meadow using an exposure of, say, one second. The resulting photograph proves to be something of a disappointment. Each firefly lights up for only a fraction of the time that it spends in flight. The rest of the time the firefly is dark, storing up energy for another flash. A one-second exposure will capture the light of only a fraction of the fireflies, those that happened to be flashing at the moment when the photo was made.

This may be our circumstance with regard to the violent galaxies. Perhaps many galaxies experience brief, recurrent episodes of violence, and the "violent" galaxies we see today are otherwise normal systems that happen to have been going through a violent phase at our epoch in cosmic history. They are fireflies that we have caught in the act of flashing, while the "normal" galaxies are fireflies biding their time between flashes.

The question remains, how do galactic fireflies flash? By what machinery can a galaxy's nucleus produce gales of energy, flinging the stuff of millions of stars into space with the abandon of passengers tossing streamers from the deck of a departing cruise ship?

Centaurus A (page 115), a giant galaxy with three times the stellar population of the Milky Way, produces great gouts of energy at many different wavelengths; alien astronomers whose eyes were fashioned to observe the world in X-rays or infrared light or radio waves would find it as commanding an object as we do in visible light. Most of these energies come from the region of the nucleus. In addition, radio energy is being generated by two pairs of clouds that lie along the galaxy's axis of rotation, that is, out from its north and south poles. Each of these pairs of radio-prominent clouds is symmetrical, one cloud on the north-pole side, the other on the south-pole side. The nearer pair lies sixteen thousand light-years to either side of the nucleus, while the other set is much farther out, at a distance of more than one million light-years from the nucleus. It is very possible that the clouds are composed of hot, thin gas that has been ejected from the nucleus.

The nucleus itself is variable, its radio and X-ray radiation altering in intensity over intervals of as little as a few days. Optically, the galaxy is very bright. At its distance of sixteen million light-years, it shines so brilliantly that it could be seen with the unaided eye in the skies of earth were it not for the thick swath of dust and gas that bisects it, cutting off our view of its central regions.

The dust ring earns Centaurus A the designation "peculiar," since normal elliptical galaxies contain little interstellar gas and dust. And while elliptical galaxies are dominated by old stars, as is the elliptical component of Centaurus A, its wreath is rich in bright blue stars, as can be seen in the photograph. The elliptical part of the galaxy glows around its edges with the red hue of older stars; in the more central regions starlight has overexposed the photograph so that colors cannot be seen there. Young blue stars are tangled through the wreath. Where we see the wreath superimposed on the inner galaxy, a coral color results from the light of the foreground blue stars mingled with that of the red stars behind.

The young stars of the wreath are very young, and must have been born near where we now find them, as none has had sufficient time to migrate far from its birthplace. The old stars of the elliptical realm are very old. Relatively few stars of intermediate age are to be found in this galaxy. It is as if Centaurus A were two galaxies merged into one. And perhaps this is a clue to solving the mystery of Centaurus A. Could it be that the elliptical component is a normal elliptical galaxy that recently swallowed a dust-laden spiral galaxy? In this scenario, the wreath is a remnant of the victim galaxy, the young stars in the wreath result from an episode of star-formation touched off by the shock of the galactic cannibalism, and the violence of the nucleus results from its having ingested interstellar material from the captured galaxy.

If Centaurus A did recently enhance its already imposing mass by absorbing another galaxy, most of the action was over by the time we came on the scene. Studies of the motions of stars in Centaurus A show that the elliptical component is rotating slowly, the wreath rotating more rapidly around it, but that otherwise the two components are drifting through space together. If a spiral galaxy was captured by Centaurus A, it has been captured for good and will never emerge.

Many of the known violent galaxies first came to the attention of astronomers using radio telescopes who noticed their unusually powerful output of energy at radio wavelengths. Radio radiation is common throughout the cosmos. Stars and even planets emit some energy at radio wavelengths, if only weakly, and normal galaxies keep up a soft radio murmur.

Most of the radio energy from a normal galaxy is produced by atoms of gas floating in interstellar space; each atom lets out a chirp of radio energy once in a great while; the number of atoms of gas in a galaxy is so large that a steady radio noise results, like the undercurrent of noise produced by a restless theater audience though only some of the people in the theater are speaking at any given time. This sort of radio noise is known as "line radiation," because each sort of atom radiates at a characteristic line, or frequency, along the radio spectrum. Hydrogen, by far the most abundant element in space, radiates at a wavelength of twenty-one centimeters; as a result, twenty-one-centimeter radio observations make for a useful way of mapping interstellar hydrogen clouds in our galaxy and in other galaxies nearby enough for this quiet but persistent form of energy to be detected.

In the far more powerful "radio" galaxies, the important source of radio noise is not the polite babble of atoms adrift, but the scream of electrons being accelerated to speeds approaching that of light. This means of energy production is known as "synchrotron" radiation, after the synchrotron accelerators used to speed subatomic particles in research laboratories. A violent galaxy may radiate one hundred times more powerfully in the radio wavelengths than does a normal galaxy, and some radio-prominent quasars (see page 177) are estimated to be pouring out energy yet a million times more powerfully still.

Cygnus A ranks as one of the most conspicuous radio sources in the sky despite its staggering distance of over half a billion light-years from earth. The radio noise comes from two lobes located symmetrically to either side of the galaxy, something like the twin radio lobes of Centaurus A. The similarity might go deeper, and the dark pinch that gives Cygnus A its hourglass appearance might be a dark wreath like the one that cuts across the face of Centaurus A.

Most radio-prominent galaxies are ellipticals, like Centaurus A and Cygnus A, but a few more nearly resemble spirals. Perseus A is radiating radio waves from a pair of lobes, but in this case the lobes are located close together near the nucleus of the galaxy , and are whirling about their common center of gravity so rapidly that they complete an orbit in only about ten thousand years. Each is estimated to be as massive as three hundred million suns. The intensity of radio output from this strange nuclear region varies acrobatically: Between 1960 and 1970 it boosted its output at the one-centimeter radio wavelength by more than five times.

The giant violent galaxy M87 heralds its presence by a bright jet that projects like a bony finger from its core. The jet is composed of hot, thin, ionized gas—what physicists call plasma—being shot from the center of the galaxy. It glows with an intense blue light produced by synchrotron radiation. This means of energy production, involving the interaction of

95,

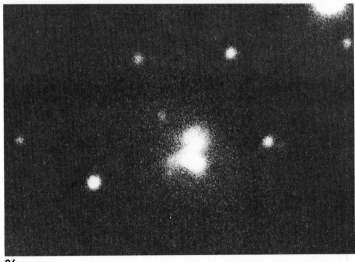

96

electrons with a magnetic field, is usually encountered as a source of cosmic radio noise, but here in M87 the gas of the jet is being propelled through the galactic magnetic field with such violence that its energy has been shifted up from radio wavelengths into the more energetic wavelengths of visible light. We can get a sense of the velocity of the jet if we consider that although it is only some fifteen thousand years old it already has achieved a length of five thousand

95, 96 The galaxies Perseus A (95) and Cygnus A (96) are powerful emitters of energy at radio wavelengths.

light-years. By galactic standards it must have appeared as suddenly as a bolt of lightning.

The jet is emerging along one of the axes of rotation of M87—from the galaxy's "north" pole, if you will—and this axis is pointed somewhat in our direction. Evidence of a counterjet extending along the opposite pole has been found, but this jet is more difficult to observe since it lies toward the opposite side of the galaxy and is rushing away from us.

This situation, too, calls to mind Centaurus A, with its two pairs of radio sources oriented along the poles as if composed of clouds of gas that had been ejected from the nucleus. It may well be that in M87 we are seeing just such a classic two-lobe radio source in the process of creation. As the twin jets of M87, slowing and dissipating, coast on into space, we may expect that their energy level will drop back into radio wavelengths, lending M87 a radio profile similar to that of Centaurus A.

The dominant galaxy of the Virgo Cluster, M87 is enthroned at the cluster's center. A substantial amount of energy at X-ray wavelengths is being generated from intergalactic space in the environs of M87, apparently produced by clouds of hot hydrogen gas. If M87 is in the habit of spitting out jets of gas repeatedly, it could have produced these clouds in past eruptions. The astronomer Iosif Shklovskii estimates that one dose of gas spewed from M87 every three thousand years would be enough to account for the intergalactic gas clouds responsible for the X-ray radiation.

M87 is massive to be sure. Its more than three thousand billion stars would suffice to populate dozens of galaxies the size of our Milky Way, itself no minor system. The gravitational force exerted by this egg-shaped aggregation of stars and its more than ten thousand attendant globular clusters is appreciable by any standard. What resides at the center of it all, at the nucleus of M87?

---

97 The giant, violent elliptical galaxy M87 (= NGC4486 = 3C274 = Virgo A) with a mass of three thousand billion times that of the sun, a halo of ten thousand globular clusters, and a protruding jet composed of knots of gas, is one of the landmark galaxies of the universe, a conversation piece for even the most cosmologically urbane.

98, 99 In color, M87 displays the candlelight hues characteristic of the old stars that dominate elliptical galaxies, while the jet glows a blue-white resulting from synchrotron radiation produced by the interaction of electrons with the galactic magnetic field. The black and white photograph, underexposed to show detail in the nuclear region of M87, reveals that the jet proceeds directly from the nucleus.

that have sent clouds of gas scudding in wildly divergent directions.

The nucleus is unusually bright both optically and in radio wavelengths. There is evidence that it ejected two clouds of material with a total mass equal to several tens of million suns along the plane of the galaxy some eighteen million years ago. It seems fair to say that M106 recently "exploded," if we understand that term to mean not that the nucleus was destroyed, but that it violently disgorged itself of material in a manner somewhat analogous to a star's generating a "planetary" nebula—that is, in a traumatic but survivable episode, like a snake's shedding its skin.

At first glance M94 (above) looks more placid than its neighbor M106, with which it shares membership in the Canes Venatici

103 A brilliant nucleus marks the violent galaxy M94 (= NGC4736).

I Cloud of galaxies. But it too proves to have been the scene of an explosion, one that may have come as recently as ten million years ago.

The otherwise unremarkable appearance of some violent galaxies, like M94, bolsters the hypothesis that violent outbursts occur episodically in normal spirals. If so, what we call violent galaxies are really normal galaxies that we happen to see during one of their episodes of violent energy production. Perhaps all spiral galaxies know how to roar, and even such apparently quiescent ones as ours and the Andromeda spiral are but sleeping beasts.

Dominated by the fixed stare of its nucleus, NGC4151 is one of the more conspicuously violent galaxies in the sky. The brilliance of the nuclear region is so overweening that it makes it difficult to discern the structure of the rest of the galaxy; observing NGC4151 is something like standing in the path of an oncoming locomotive at night and trying to make out the shape of the locomotive behind its headlamp.

Observers who have managed to overcome this difficulty have found the structure of the galaxy to be that of a weakly barred spiral. The bar is not visible in the photograph (right), having been swamped by light from the nucleus. If we could "turn down" the nucleus we would see that the galaxy resembles NGC4156, the barred spiral that lies off to one side of it near the edge of the frame. (Although the two galaxies appear close together in the sky, NGC4156, at a distance of four hundred forty million light-years, is nearly seven times farther from earth.)

While the bar region of barred spirals normally is inhabited primarily by old red giant stars, the bar of NGC4151 shines most brightly not in red light but in blue and ultraviolet. Perhaps this blue light is coming from new stars that formed out of raw gas blasted from the nucleus of the galaxy into the spiral arms. The bars might be acting as conduits for material ejected from the nucleus, and might have become studded with new stars along the way in a manner analogous to the way mineral deposits build up on the inside of a water pipe—though the galactic bar is not, of course, a solid object like a pipe, but is rather an association of gas, dust and many stars.

The brightness of the nucleus of NGC4151 is variable. Soviet astronomers have identified what they believe to be a one-hundred-thirty-day period of variability for the nucleus, with a seventy-day pulse imposed upon it. One of the mysteries of violent galactic nuclei is how something that we would expect to be rather large manages to go through these gyrations, antic as a bee dance.

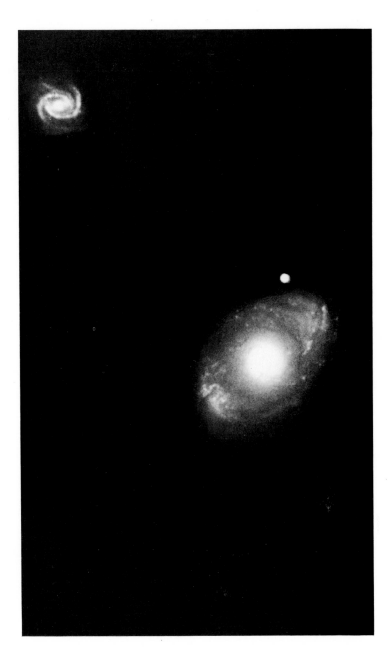

104 Obscured by the glare of its brilliant nucleus, NGC4151 possesses a barred structure similar to that of the background galaxy seen near the edge of the frame. Most of the foreground stars have been removed from this photograph so that the galaxy may be seen as in its natural habitat, floating in starless space.

If some galaxies may be said to be exploding, NGC253 may be said to be simmering. Clouds of gas are wafting outward from its central regions at a robust though not apocalyptic rate. Even so, the nucleus is losing gas rapidly enough that had it been simmering continuously throughout the history of the galaxy it would have depleted itself long ago. Since it has not done so, we may conclude either that the nucleus has acquired fresh material, perhaps by sucking in an intergalactic cloud, or that the outflow is intermittent.

A giant spiral, NGC253 is roughly the size of the Andromeda Galaxy. It belongs to the Sculptor Group, nearest neighbor to the Local Group, at a distance of a little over ten million light-years.

105, 106 The large spiral NGC253 shows only mild signs of activity in its nucleus.

107

108

"Peculiar" galaxies are those that fail to fit into a category under existing human schemes of galaxy classification. The term serves less as a description than as a convenient basket into which to dump galaxies our understanding of which is so limited that we cannot yet accomplish the zoologist's task of naming their family. Until we understand galaxies well enough to replace the term with a more learned one, we shall continue to enjoy the spectacle of humans calling galaxies "peculiar."

Many of the peculiar galaxies are violent, or are found to be conspicuously interacting with nearby galaxies, or both. NGC2146 is a good example; violently perturbed in appearance, it is a powerful source of radio energy.

NGC2685 arouses interest in that its structure is exceedingly unusual without there being anything else noticeably odd about it. The central component resembles an SO galaxy viewed nearly edge-on, wrapped in a set of gigantic hoops oriented perpendicular to its plane. A faint, barely visible outer ring encircles the entire system. Possibly we are seeing here the collision of two galaxies.

107, 108 These two strange-looking galaxies are classified "peculiar." NGC2146 (**107**) is classified SAb (pec), while NGC2685 (= Arp336) (**108**) is listed SO (pec).

IV/Interacting Galaxies

How is it that the sky feeds the stars?

—LUCRETIUS

A Journey between Interacting Galaxies

A high point in our intergalactic journey comes when we steer our ship between a pair of interacting galaxies. We have chosen to fly through the relatively narrow corridor separating two major spirals. They constitute a binary system, two galaxies bound together gravitationally as are the Milky Way and the Andromeda Galaxy. For most of their history they have stayed well apart, but now they are passing within only a couple of galactic diameters of each other, and it is at this dramatic stage in their interaction that we are to come between them.

The first mate is nervous. He points out that if we could see where we are headed in terms of Einstein's space-time continuum, we would perceive that our course lies along a precariously narrow ridge between two enormous wells created by the gravitational potential of the two galaxies. "We are steering between Scylla and Charybdis," he warns, "or rather Charybdis and Charybdis, for we'll have whirlpools to either side."

The galaxies are passing at an orientation that brings them almost face-on to each other, like a pair of cymbals. From a distance we see them edge-on. As the months pass and we draw closer, our perspective makes them appear to swing open like a pair of doors. The doors do not open evenly, but remain somewhat closer together near the top, where the relative inclination of the galaxies to each other has brought them closest together. Luminous tendrils bridge the gap between them there. Soon we will be amid the spectacle.

Thin clouds of hydrogen gas pervade the intergalactic space surrounding the two galaxies, and as we plunge through these their friction produces a sustained high-pitched wail from the ship's hull. To ward off nervousness we take solace in determinism: It is comforting to reflect that the cymbals cannot choose to clash when we pass between them, but will continue to follow the orbits dictated by Newton's and Einstein's laws. We have watched the projected course of the pair many times in computer simulations—a do-si-do, the two galaxies turning sharply around their common center of gravity and then parting a hundred million years from now. We ought to be able to sail between them without mishap. Still, we keep a cautious eye on our course; none among us wants us to attempt to be the first to fly at nearly the speed of light through a spiral galaxy edgewise.

To further comfort ourselves we discuss interacting galaxies in general. We remind ourselves that they are not rare. All galaxies, we reassure one another, may be said to be interacting, in that all respond to the general gravitational field of the universe, to which millions of galaxies contribute.

Didn't all the galaxies come from an undifferentiated soup of matter that permeated the universe long ago? And wasn't their formation a story of vortexes arising from the primordial soup, condensing to form the pairs, groups and clusters of galaxies we see today? And isn't the structure of galaxies, which we find so lovely a sight, but the visible message written by the invisible hand of these gravitational interactions?

There have been many close interactions of galaxies in the past, some perhaps involving the Milky Way and the Andromeda spiral, and the galaxies survived them in good order. They were merely twisted, their disks distended, their nuclei banked into fire, millions of their stars blasted into space.... Whole galaxies wrenched out of shape....

We fall silent. The ship's hull moans.

Ultimately we find ourselves between them. One spiral galaxy hangs to port, the other to starboard, two celestial wheels, ourselves at the axle. Their starlight flooding through the ports bathes the interior of our ship in a light such as none before us has known.

We view the spectacle from the overhead observation room, a transparent bubble that the ship's designers whimsically modeled after the domed railroad passenger cars once popular in North America. We turn out the lights and look above us to view the parts of the spirals where their mutual inclination has brought the disks closest together.

There the intergalactic gap is bridged by luminous tendrils that hang far above us like vines in an arbor. We can see that they are composed of gas and millions of stars being stripped from the lesser of the spiral galaxies and transferred to the more massive.

"A stellar caravan," remarks the first mate. "'The dogs bark; the caravan moves on.'"

We gag the first mate with an antimacassar, and resume watching, in silence, the transactions of galaxies.

The stars of a globular cluster flash past at close quarters, frightening us all. Amid screams, someone thinks to ungag the first mate. He bids us to be not afraid. Our course is taking us through the outskirts of one of the globular clusters that belongs to the halo of one of the galaxies, he shouts. Stars are flashing by the windows like balls from a Roman candle. He had intended to warn us, he shouts.

Still, we are quick to descend the ladder. It is days before anyone goes back up there.

Weeks pass and the twin galaxies crawl away aft. We welcome the sight of the dark intergalactic spaces we once had feared.

IV/INTERACTING GALAXIES

The M51—NGC5195 System

The story of interacting galaxies—and indeed the story of galaxies themselves and the stars, planets and interstellar clouds that make up galaxies—is largely a story of gravitation. This attraction of matter for matter holds together the atoms of stars and planets, maintains the congregations we call galaxies, and binds the galaxies together in groups, clusters and superclusters. Subtract gravitation, and the universe would explode into vapor.

So ubiquitous is the influence of gravitation that one can go through life scarcely noticing it, as a fish may devote little thought to water. Yet, precisely because gravitation is ubiquitous, one who grasps its essentials can predict with accuracy an enormously wide range of physical phenomena all over the cosmos. In our world this insight befell young Isaac Newton, who intuited that the force responsible for the apple's fall to earth might also account for the orbits of astronomical bodies. Newton's equations came close enough to the truth that they are still used today to account for the dynamics of interacting galaxies, where the effects of gravity generate the abnormal and so render themselves conspicuous to our eye.

The normal gravitational milieu of a major galaxy might be described as democratic and sovereign. It is democratic in that the form of a galaxy reflects a gravitational condition in which every gram of matter belonging to the galaxy gets its vote. The great majority of this matter is bound up in stars. The highest concentration of stars is found in the central part of the galaxy, with stellar densities falling off as we look to the outer regions. The orbit that each star follows traces out its response to the gravitational environment created by all its compatriot stars.

In the case of our sun, an outer disk star in a spiral galaxy, the orbit takes the form of a gentle ellipse, focussed on the center of the Milky Way Galaxy and tilted slightly with respect to the plane of the galaxy. The sun completes one orbit every two hundred fifty million years. Most of the stars in our galaxy lie inside the sun's orbit, concentrated toward the galactic center, so that we can approximate the sun's orbit mathematically by pretending that it is orbiting a point at the center of the galaxy whose mass is equal to that of many billions of stars. In Newtonian terms, we say that the sun is responding to the gravitational "force" exerted by all the other stars; in Einsteinian terms, we say that its orbit represents a "geodesic," meaning that it is pursuing the most efficient course available to it across the contours of the space-time continuum. The result of all this is what we see as a normal galaxy—a generally relaxed, approximately symmetrical, reasonably orderly congress of stars, secure within a gravitational arrangement, worked out, as it were, among themselves.

Galactic sovereignty pertains insofar as a galaxy's gravitational milieu remains predominantly its own. Normally each major galaxy is free to reign within a comfortable volume of space, its neighboring large galaxies keeping their distance. Sovereignty is violated when a neighboring galaxy drifts close enough that its gravitational dominion begins to interfere with domestic order. Gravitational force decreases by the square of the distance, so an equal degree of disturbance may be created by a massive galaxy that passes at a distance or by a less massive galaxy that comes quite close or collides.

The effects of gravitational interference upon the structure of the galaxies involved can be dramatic: One such dramatic encounter in our cosmic neighborhood involves M51 and NGC5195. Both are spirals, each about one-half the mass of the Milky Way. M51 appears nearly face-on while its companion, viewed nearly edge-on and badly distorted by the effects of the interaction of the two galaxies, is usually classified as an irregular.

M51 is one of the most achingly beautiful spirals known to humankind, but its beauty contains an element of distress, like that of a racehorse or a bonsai tree. It is in fact a markedly disturbed galaxy. The spiral arm nearer to the companion galaxy reaches beseechingly after it, at one point tucking under an inner arm, while on the opposite side the arm has sprung well away from its normal pitch.

---

109 The conspicuous spiral galaxy M51 (= NGC5194) and its companion NGC5195 passed close to each other during the last several hundred million years; effects of the encounter may be seen in the elongated shape of the disk of M51 and the sprung pitch angle of its spiral arms.

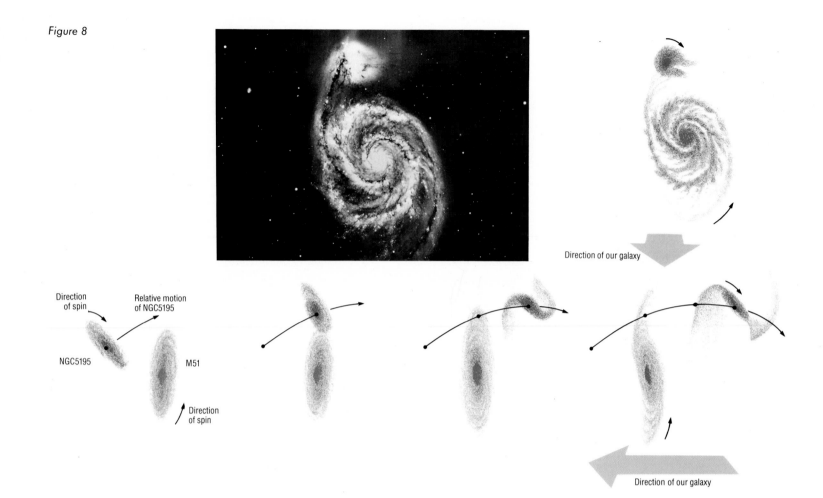

Figure 8

Direction of spin

Relative motion of NGC5195

NGC5195

M51

Direction of spin

Direction of our galaxy

Direction of our galaxy

A computer-generated reenactment of the encounter that produced these distortions, as seen in the illustration (page 131), indicates that NGC5195 has followed a boomerang trajectory that carried it past M51 millions of years ago. Now it is on the far side of M51 and is moving away, leaving ample evidence of disturbance in its wake. The normally almost circular disk of M51 has been squeezed into an ellipse by the gravitational pull of NGC5195; part of the disk nearest to the companion galaxy was tugged after it, while the part of the disk on the opposite side—responding to a decreased gravitational potential created by the fact that much of the other side of the galaxy had been removed to a greater distance—expanded, leaving the arms on both sides loosened like a relaxed mainspring.

The encounter of these two galaxies stripped millions of stars from their parent systems and left them adrift in intergalactic space. Had we evolved on a planet circling one

such ejected star, we would find our night sky all but empty of stars, punctuated by only a few stars with which we shared our habitat in exile. Dominating the night sky would be the two great galaxies—the shattered disk of NGC5195 in one direction, in the other the great wheel of M51. Astronomers deciphering the relative motions of the two galaxies and of our home stars could piece together the story of where we had been millions of years earlier, and how we had been orphaned.

Figure 8. M51–NGC5195 Interaction
Both M51 and its companion NGC5195 were badly distorted by a recent close encounter, as indicated in this "side view" that reconstructs their history over the past few hundred million years. NGC5195 is proceeding away from M51, but since it lies behind M51 as seen from the Milky Way the two galaxies look to us as if they were still entangled.

132

110

110 In color, M51 exhibits fields of young blue stars along the spiral arm nearer to the companion galaxy NGC5195; possibly they were created as a result of disruptions in the interstellar clouds along that arm produced by the gravitational interference of the passing galaxy.

The M81—M82 System

112

113

Another case of galactic interaction that occurred recently in cosmic history may be found in the M81 group, a neighbor of the Local Group only some ten million light-years away. The flagship of the group, M81 is nearly as large and populous as the Milky Way.

One hundred thousand light-years away from M81 lurks M82, a mysterious galaxy that has played the role of the Sphinx in contemporary astrophysics. It is probably a spiral viewed nearly edge-on; in any case it looks most odd. Its central regions are mottled in appearance, with dark clouds of remarkable extent silhouetted against massed starlight. Here are to be found bright nebulae in such abundance that if our solar system were located there the sky would be a glowing crazy quilt. Much of the interstellar material, dark and bright nebulae alike, extends far out of the plane of the disk, as if M82 had been shaken by some sort of galaxy-quake. Crowning its eccentric appearance are a pair of what appear to be clouds of hydrogen gas that protrude from the galactic poles.

A host of theories has been presented to account for what is going on within the shrouded confines of M82. Some of them are as imaginative as those that in times past promoted narwhales to the status of mermaids; these seem in retrospect to have resulted as much from the predispositions of the theorists as from evidence offered by the galaxy itself. Like the Sphinx, M82 has conversed with us less in answers than in echoes.

Still it is possible to reconstruct a likely history of M82: Some two hundred million years ago it was churning peacefully through space, a small spiral going about its business. Then the enormous M81, ten times more massive than M82, swept past like an ocean liner past a sailboat. The gravitational wake of the larger galaxy washed across the smaller one, altering the orbits of millions of its stars and shocking its interstellar clouds into collapse, producing millions of new stars. Much additional interstellar material was jolted out of the plane of the galaxy, either tugged away by the pull of the receding M81 or blasted out of the disk by the many

---

111 M81 (=NGC3031) is the dominant spiral of a group of galaxies located only ten million light-years away.

112 A wide-angle view shows the relationship in the sky of M81, M82 (=NGC3034), and a dwarf satellite of M81, NGC3077, visible at the corner of the frame. Studies of the relative motions of the galaxies indicate that M81 and M82 passed each other two hundred million years ago and are now drawing apart.

113 A radio map of the system shows that M81 and M82 are wrapped within a common envelope of intergalactic gas.

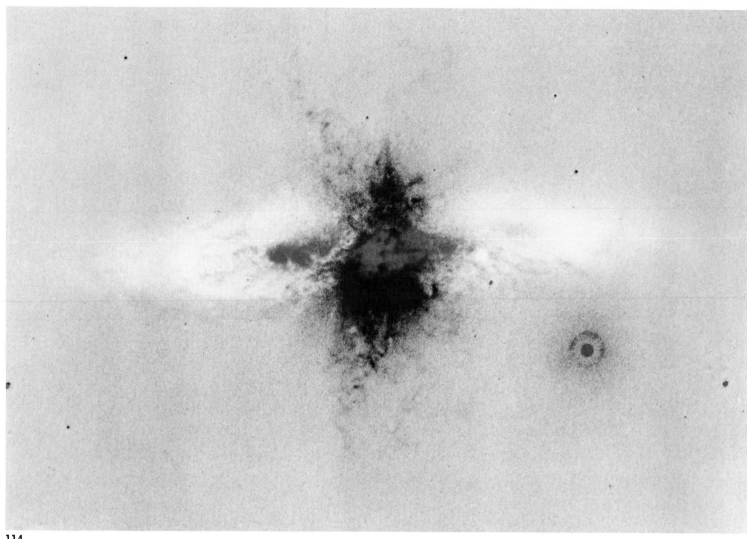

114

supernovae that characteristically flare up among young massive stars. The displaced material subsequently succumbed to the gravitational attraction of its parent galaxy and fell back into it, touching off another episode of star formation. Astrophysicists estimate that the first epoch of extraordinary star formation occurred primarily in the disk forty million years ago, and that the more recent one, in the central regions, is continuing today.

If this reconstruction is even approximately correct, then the M81—M82 system adds evidence in support of our dawning perception that some of the more spectacular events that occur in galaxies may result from their interactions with other galaxies.

114 Photographed in a wavelength of red light produced by hydrogen gas and designated Hydrogen Alpha, the clouds protruding from the central regions of M82 reach as much as ten thousand light-years into intergalactic space.

115 M82, highly disturbed in appearance, appears to be a small spiral galaxy seen nearly edge-on whose interstellar material has been shocked well out of the plane of the disk by the gravitational tug of the passing M81.

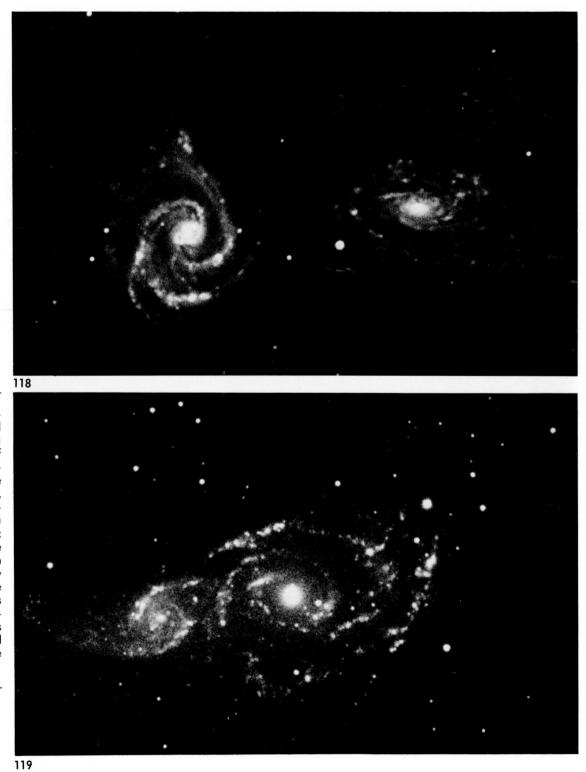

118–121 Some intergalactic encounters, like those of the spiral galaxies NGC5426/27 (**118**) and NGC4567/68 (**120**) and of NGC2207/IC2163 (**119**), produce only mild disruptions in the structure of the galaxies involved, while others, like those of Stephan's Quintet (= NGC 7317–20) (**121**) look profoundly dynamic and perturbing. All, however, are changing so slowly by human standards as to appear virtually eternal; had our predecessors the Cro-Magnons built telescopes and photographed these galaxies, their photographs, taken tens of thousands of years ago, would be virtually indistinguishable from these.

118

119

140

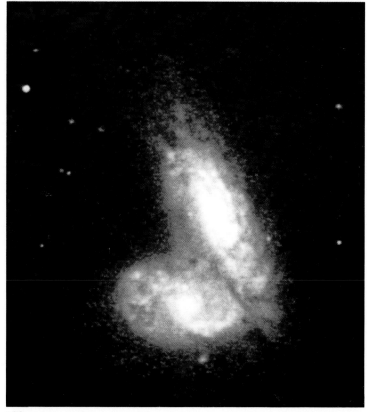

120

Degrees of Interaction

The dance of interacting galaxies can be as stately as a minuet or as frantic as a mazurka. The pairs of spiral galaxies seen here are interacting in stately style, their unsprung arms waving a gentle semaphore.

In contrast, the galaxies of Stephan's Quintet (right) are deeply intertwined and bear the stamp of haste. The placid-looking spiral to the lower left of the other four is in the foreground and does not belong to the group, which ought therefore to be called Stephan's Quartet. Some observers have found evidence that the galaxies of the quartet are in the process of merging into a single galaxy, while others have reached the opposite conclusion and predict that the quartet is destined to fly apart.

One way or another, the issue has already been resolved. The galaxies have either merged or parted company. The three hundred million light-years worth of light stretching from the quartet to our galaxy contains the news of their fate.

121

"Rattail" Galaxies

122

Galaxies that nearly collide can form long tails full of stars and interstellar material that extend for several galactic diameters into space. The sequence of events that makes this possible appears to be the following:

Two galaxies of approximately equal mass—normally a binary pair traveling in highly elliptical orbits that only occasionally bring them into close proximity—pass close together. On the sides nearer each other, billions of stars are torn from their orbits, many to be left behind in intergalactic space. As a result of this depletion of mass, each galaxy loses some of the gravitational attraction that had been holding it together, and stars on the far sides are permitted to wander out into intergalactic space. It is as if one were whirling a bola overhead and suddenly fed it more string, permitting it to fly outward. The result of this occurring in each of the encounter galaxies is the pair of tails we see.

It takes some time for the tails to unwind themselves across such great distances; by the time they have manifested themselves fully, the encounter is long past and the galaxies involved are well on their way to regaining a relatively untroubled appearance. In a few hundred million years all will be calm and it will be difficult to tell that anything spectacular happened here.

122 The "rattail" galaxies NGC4038/39 (above) and NGC2623 (page 143) sport long plumes of ejected stars and interstellar material generated by the gravitational interaction of these two galaxies of approximately equal mass.

Figure 9 "Rattail" Galaxies.
A close encounter of two galaxies of approximately equal size upsets the internal balance of both, flinging billions of stars far into space in a pair of long arms or "rattails." This depiction is based on a computer-generated reconstruction of the sequence of events.

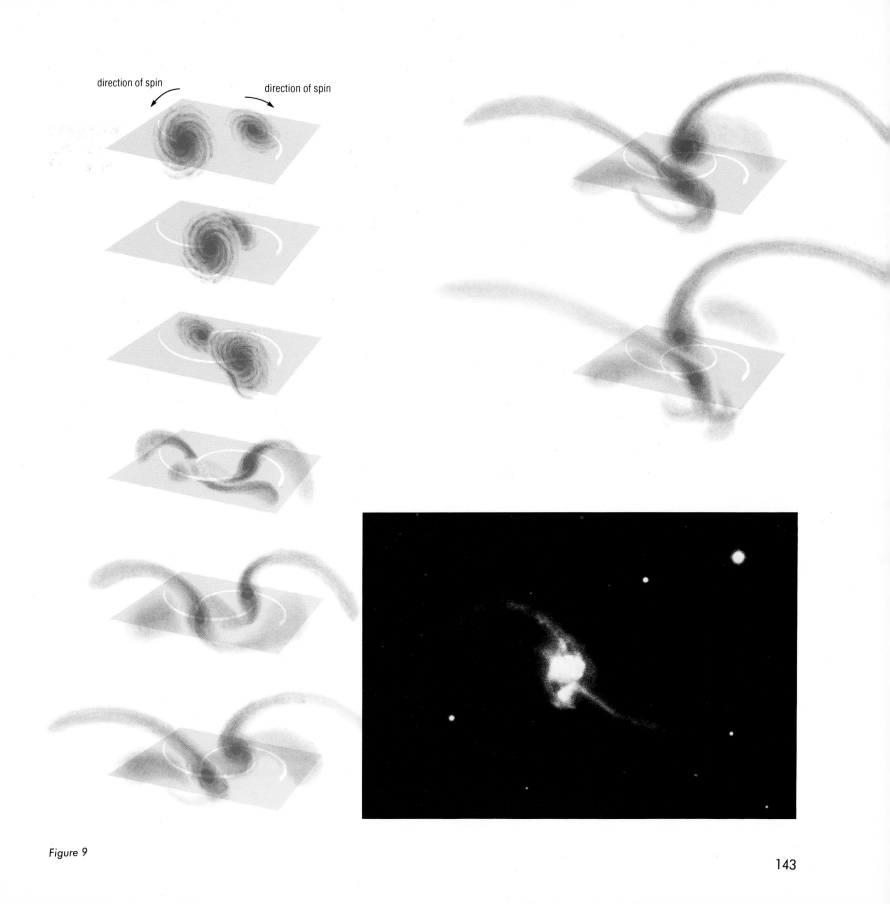

direction of spin   direction of spin

Figure 9

143

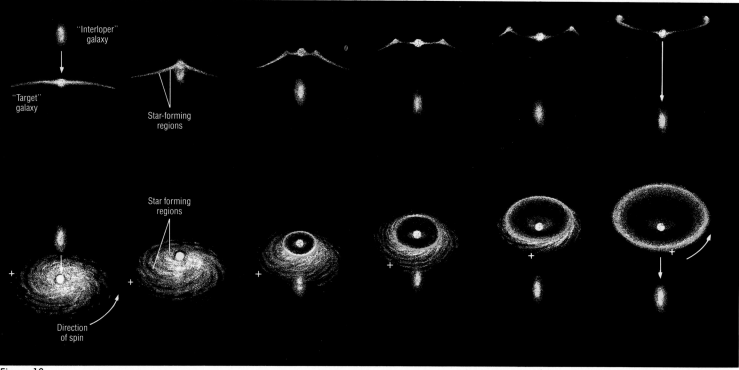

Figure 10

Ring Galaxies

Ring galaxies are created when a large spiral collides with a small galaxy or an intergalactic gas cloud. The ring structure, a temporary affair, results from the profound alteration of the gravitational field of the large galaxy that occurs when the interloper blunders into it.

Let us imagine that an interloper galaxy strikes a large spiral more or less dead center. Billions of interloper stars drift among those of the central region of the spiral, passing among them roughly at right angles to the plane of the larger galaxy. Interstellar space is roomy, and few if any stars collide, but the temporary occupation of the spiral galaxy by billions of alien stars has the effect of greatly enhancing the local gravitational attraction. Disk stars are drawn inward by the increased gravitation.

But by the time many of the stars arrive in the central regions, they find that the show is over. The intruder galaxy has passed on, taking its gravitational potential with it.

Released from the force that had attracted them, the disk stars rebound outward in an expanding ring. The shock of all this turmoil engenders the wholesale collapse of interstellar

Figure 10. Ring Galaxies
The creation of a ring galaxy is here reconstructed in two perspectives. A small "interloper" galaxy passes through or near the center of a large spiral, drawing the stars and interstellar material of the large spiral toward the interloper. When the interloper departs, the stars, gas and dust are released to fly back outward as a ring, spangled with the light of new stars set off by the shock of the event. This disfigured galaxy will soon regain its normal appearance.

123 An interloper galaxy speeds away, trailing a plume of gas, dust and stars and leaving behind a ring galaxy that already is beginning to reorganize itself into a normal spiral; the ring galaxy is catalogued as Object RG33 No. 754.

123

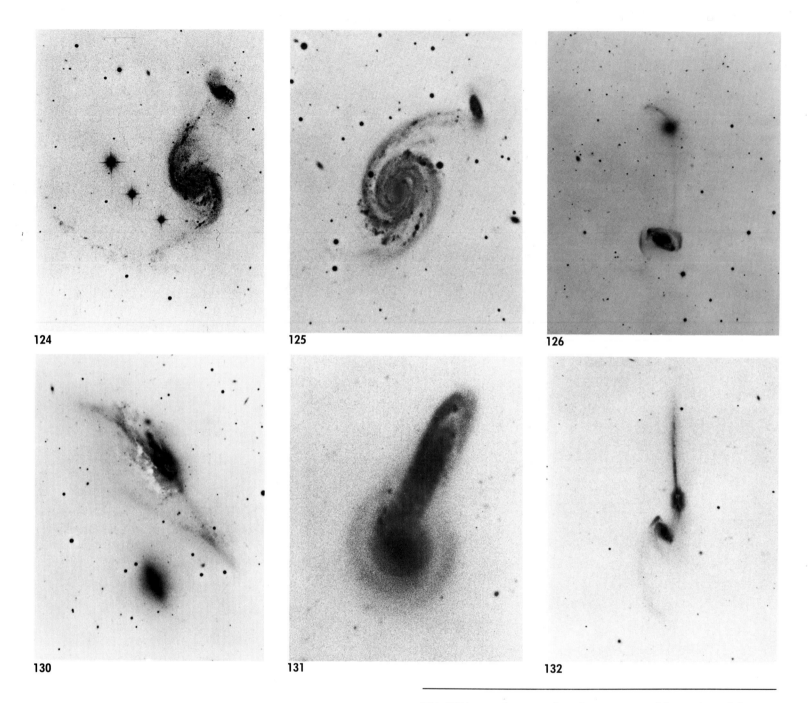

124

125

126

130

131

132

124–135 Interaction can give rise to an exquisite variety of form in galaxies. Here negative prints are employed, to show in greater detail in the rarified outer regions of each galaxy. They are: **124**: NGC2535/36 (= Arp 82); **125**: NGC7753/52 (= Arp 86); **126**: NGC5216/18 (= Arp

146

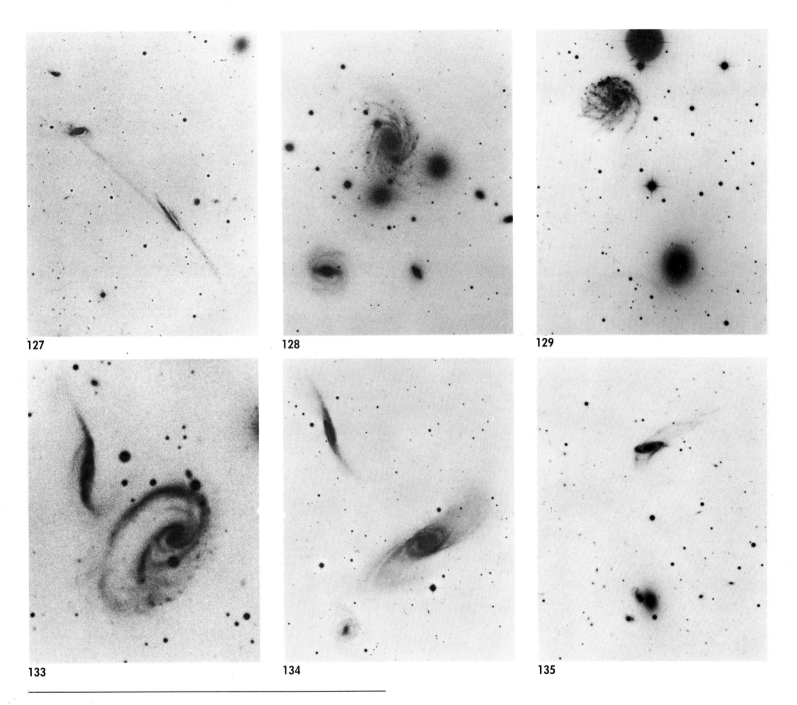

127

128

129

133

134

135

104); **127**: IC1505 (= Arp 295); **128**: NGC70 (= Arp 113); **129**: NGC2275/2300 (= Arp 114); **130**: NGC4438 (= Arp 120); **131**: NGC5544/45 (= Arp 199); **132**: NGC4676 (= Arp 242); **133**: Arp 273; **134**: NGC5566/60/69 (= Arp 286); **135**: NGC5221/22/26 (= Arp 288).

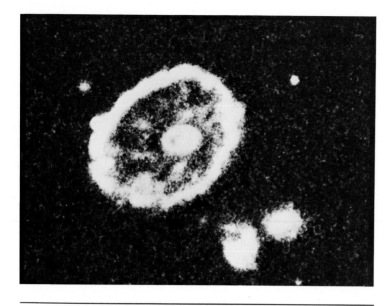

136 The Cartwheel is believed to be an otherwise normal spiral that has been violently disturbed by a direct collision with a companion galaxy, probably the upper of the two dwarf galaxies seen below and to the right of the Cartwheel.

clouds to form new stars. The expanding ring scintillates as it expands, twinkling with fresh starlight.

The spectacle is fleeting. The ring cannot long endure. Its stars sort themselves out into relaxed orbits and the galaxy resumes a normal appearance. Soon the only evidence of the collision is the spiral's unusually bright nucleus. Feeding on gas stripped from the intruder galaxy, the nucleus glows much like that of a Seyfert galaxy (see pages 117, 120, and 123).

The Cartwheel ring galaxy (left) may look insubstantial as a smoke ring, but it embraces as much space as the Milky Way, and harbors more stars. The companion galaxy believed to have collided with it is the slightly more distant of the two visible in the photo, the one that lacks a comma-like spiral hook. Measurements of its velocity lead to the conclusion that it passed through the center of the Cartwheel two hundred fifty million years ago. This agrees well with estimates of the age of the ring, adduced by measuring its rate of expansion and running the rate backward to the time when it would have been compressed near the nucleus, roughly three hundred million years ago. For almost every known ring galaxy a companion galaxy can be found nearby, slinking away from the scene of the collision.

V/Clusters of Galaxies

*In the universe the difficult things
are done as if they were easy.*

—LAO-TSU

A Journey through the Local Supercluster

Our old ship has gone far. We have edged so close to the speed of light that sometimes we feel we have become like light, fleet and insubstantial, velocity itself our only home. Decades have passed on board. There have been deaths and births, happiness and sadness, success and failure—in short, decades of life. The string quartet broke up years ago. The cook has grown grumpy from the ebbing of both praise and blame. Scholars complain about the limitations of the ship's vast library. We who set out on this journey when so young have become the elders. Occasionally we talk of putting in at a planet like earth, near a star like the sun, in a galaxy like the Milky Way, there to make a new start. But we are going so fast that just to decelerate would be the work of many years. So we fly on, like light.

Where previously we studied the form of galaxies, now our attention is drawn to the form of clusters of galaxies. Here we find order, intelligibility, a deep coherence underlying the diversity of the universe.

Clusters of galaxies, we see, display varieties of forms within a general pattern. The most straightforward way to arrange them is along a continuum in terms of structure, with the most regular clusters of galaxies toward one end and the most seemingly chaotic toward the other. The regular clusters are spherical or elliptical in shape, their galaxies concentrated at the center of the cluster. The irregular clusters, at the other end of our spectrum, are shambling and clumpy, often taking the form of extended chains of galaxies; they show little or no concentration toward the center. Intermediate between the two extremes are clusters that display some of the characteristics of both regular and irregular clusters; in some instances these consist of a central elliptical concentration surrounded by a halo or disk of more thinly distributed galaxies.

The forms of the clusters unavoidably call to mind the analogous forms of galaxies themselves: To some degree spherical clusters resemble spherical galaxies, irregular clusters resemble irregular galaxies, and intermediate clusters are not wholly unlike spiral galaxies, with their mixture of characteristics of both types. Our curiosity about this parallel deepens when we learn that the sort of galaxies predominant in each cluster is closely related to the form of the cluster itself. The spherical clusters have the largest plurality of elliptical and SO galaxies, while irregular clusters are dominated by spirals and have relatively few ellipticals and SOs. And what spiral galaxies there are in spherical clusters tend to be segregated toward the outer regions, or halo, of the cluster—much as globular star clusters occupy halos surrounding elliptical galaxies. The evidence seems compelling that the form taken on by a galaxy cannot have been determined solely by forces internal to that galaxy, but must reflect something of the milieu of the cluster to which it belongs.

Having taken this step up the hierarchical ladder, we are inclined to take an additional step and inquire whether clusters of galaxies belong to still larger associations. Here again our curiosity is rewarded. Many of the clusters prove to be members of superclusters—clusters of clusters of galaxies.

Clusters of galaxies typically occupy volumes of space with diameters of roughly thirty or forty million light-years. The diameters of superclusters are ten times greater, on the order of three hundred to four hundred million light-years. Even on this scale we find evidence of order and consistency. Some of the superclusters consist of a central zone where clusters of galaxies are relatively concentrated, surrounded by a flattened disk of more thinly distributed clusters, an arrangement at least faintly reminiscent of the structure of spiral galaxies. And possibly superclusters too are rotating.

Now when we look back to our home galaxy we may view it in a supergalactic context. The Local Group is a small cluster of galaxies located in the outskirts of the Local Supercluster. Many of our neighboring small clusters—the M81 Group, the M101 Group, the Sculptor Group—are likewise members of the Local Supercluster. The supercluster consists of a concentrated core, designated the Virgo Cluster, and an extended halo to which the Local Group and its neighboring groups belong.

A few of us gather in the observation dome after dinner. We trace for one another the structure of the Local Supercluster spread out before us, as once long ago we mapped the disk of our home galaxy. We talk of the old mystery that closes the circle of life—that the incomprehensible thing about nature, in Einstein's phrase, is our ability to comprehend it. However fast we go or far we travel, we have not fled one micron from this mystery. We can feel its breath, touch its face; it is our breath, our face.

The first mate stands and quotes a sentence from Carl Friedrich von Weizsäcker, a physicist and philosopher of science who lived millions of years ago on earth. "All our thinking about nature must necessarily move in circles or spirals; for we can only understand nature if we think about her, and we can only think because our brain is built in accordance with nature's laws."

The captain runs a hand through his white hair.

"Spirals," he says. "Our thinking expands as it circles. It moves in spirals."

V/CLUSTERS OF GALAXIES

The Form and Variety of Clusters and Superclusters

A cluster of galaxies may be defined as an association whose galaxies are bound together gravitationally. The shape of the orbit of each galaxy is determined by its gravitational environment in the cluster. In the case of a loosely organized cluster, the orbits of the galaxies might resemble the easy loops along which stars move in open star clusters, while in the denser spherical clusters of galaxies, the galactic orbits more nearly resemble the tighter orbits of stars in globular star clusters. In the Local Group, a small, loosely organized cluster of galaxies, the fundamental structure is binary—two galaxies, ours and the Andromeda Galaxy, constitute most of the mass of the cluster and orbit their common center of gravity.

Clusters of galaxies are found in a variety of forms that vaguely resemble the various forms of galaxies themselves. The so-called spherical clusters occupy a roughly spherical—elliptical would be a more accurate term—volume of space, their galaxies concentrated toward the core and scattered at the outskirts. This structure is reminiscent of globular star clusters and of elliptical galaxies, though the difference in scale is appreciable. If we were to represent a globular star cluster by a dot the size of a period on this page, a spherical galaxy would be a sphere seventeen feet in diameter, and a spherical cluster of galaxies would be the better part of a mile in diameter. Irregular clusters of galaxies, as their name implies, resemble giant irregular galaxies in their almost chaotic form. Many clusters occupy an intermediate position; in some cases these are found to consist of an elliptical aggregation at the center with a surrounding halo which, when flattened, resembles the structure of a spiral galaxy.

The size of regular and intermediate clusters of galaxies is normally a few tens of millions of light-years, while irregular clusters, like irregular galaxies, are much more varied in size, and include in their number many dwarfs. The Local Group, a few million light-years in diameter, is probably best classified as a dwarf irregular cluster of galaxies.

The types of galaxies found in clusters—and the vast majority of galaxies belong to clusters—reflect the structure of the cluster. Spherical clusters abound with ellipticals and their cousins, the SO galaxies. Irregular clusters are composed largely of spirals, with only a few ellipticals in residence. Intermediate clusters are populated by a mixture of types of galaxies.

This connection between the nature of clusters and the nature of the galaxies that inhabit them argues persuasively that the clusters are primordial, that they represent vast tracts of matter that were portioned out early in the history of the universe, before the galaxies began to form or at least before they had had time for the process of their formation to proceed very far. There must be a strong hereditary component—by which I mean such parameters as how much mass went into a protogalaxy, and what was the temperature and density of that mass—involved in the determination of whether a galaxy ends up as an elliptical, a spiral or whatever, and this hereditary influence ought to be traceable to conditions pertaining in the cluster in which each galaxy was born.

However there is also evidence for a strong environmental component acting upon galaxies throughout their lives. The experiences that befall a galaxy in the course of billions of years will differ considerably according to the nature of the cluster to which it belongs. A galaxy belonging to a loose irregular cluster follows a lazy orbit that only occasionally will bring it close to another major galaxy, and so it will have to endure galactic collisions rarely if at all. The environment is much different for a galaxy in a spherical cluster whose orbit plunges it into the densely populated cluster core. There it will be plucked at gravitationally by passing galaxies like a wealthy tourist pushing through a street of beggars, and collisions will befall it every billion years or so. A galaxy that passes directly through the center of a spherical cluster may find the conflicting gravitational pull of the thousands of galaxies surrounding it sufficient to tear it to pieces.

Supergiant galaxies are found at the center of many spherical clusters, and may well owe their opulence to their

---

Figure 11. Nearby Groups of Galaxies
Most galaxies belong to groups. Here are plotted a number of the groups of galaxies that have been identified in our part of the cosmos. The plane of the map is the "supergalactic plane" of the Local Supercluster of galaxies. The groups of galaxies, though typically irregular in form, are for simplicity depicted as spherical volumes of space. The numbers within each sphere indicate the distance of the cluster below (negative values) or above (positive values) the supergalactic plane. The concentric circles designate distances on the plane from the center of the Local Group. All these distances are in millions of light years, and all should be regarded as approximate.

Figure 11

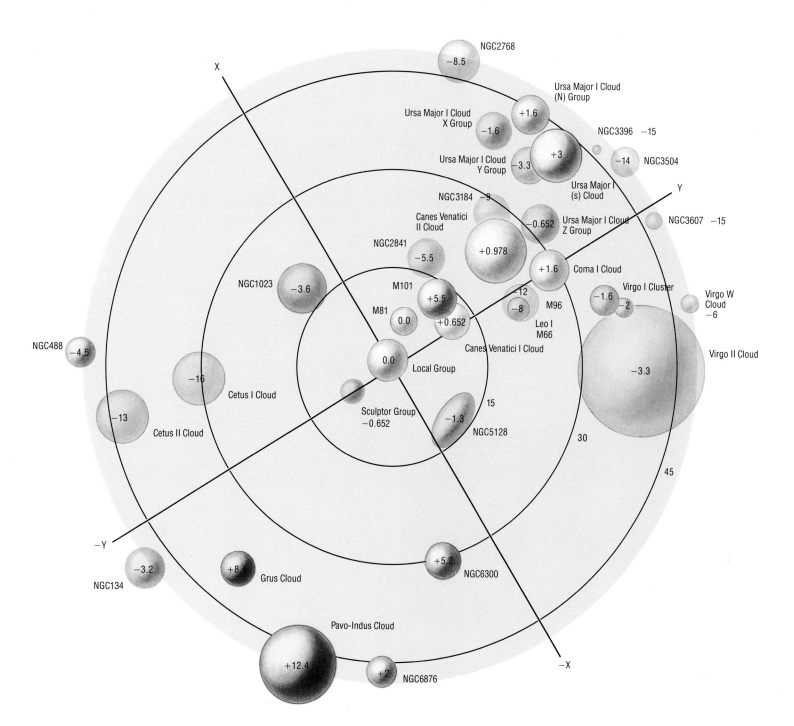

cannibalizing galaxies that blundered into the center of the cluster and fell victim to them. These supergiant core galaxies typically combine characteristics of the various sorts of galaxies they consumed—they may for example combine the shape of ellipticals with the rich interstellar clouds characteristic of spirals. And often they are found to have multiple nuclei, which may betray their predatory history as do the beaks of squid found in the bellies of whales.

Many clusters of galaxies in turn belong to superclusters, and the superclusters show evidence of structural order and variety analogous to that of clusters and galaxies. The fact that superclusters can be identified and that they bear the stamp of familiar physical laws offers reason to hope that one day we will be able, with reasonable accuracy, to reconstruct how they were formed and to predict their futures. So adept a level of comprehension might be expected to engender many a beautiful image in the mind's eye. For one thing, we might be able to correctly envision the dynamic behavior of superclusters with regard to the expansion of the universe.

"Expansion" is a way of saying that the clusters of galaxies are moving apart from one another at a velocity proportional to their distances apart from one another. Gravitationally bound, the clusters do not themselves expand. Rather, the expansion of the universe takes place out between the clusters. The clusters are departing from one another like swarms of bees that start out from one forest but diverge along differing paths of migration.

What about the superclusters? The expansion of the universe comes to predominate over local gravitational attraction on scales of something over three hundred million light-years, which is just about the average diameter of a supercluster. So it would seem that the gravitational interaction of superclusters must produce awesome distortions and transformations in supercluster structure as the ballooning of the universe tugs them apart.

When we can reconstruct the behavior of, say, the Local Supercluster over a period of billions of years, will we find that its motion more nearly resembles the orderly spin of a spiral galaxy or the gentle sway of sea grass in an ocean current? Or may its behavior, as is often the case in physics, evidence qualities that we had not anticipated, working more to engender fresh human metaphor than to lend itself to the metaphors of old?

The Virgo and Coma Clusters

137

The Virgo Cluster (right) lies at or near the center of the Local Supercluster, while our Local Group occupies a position toward the outskirts of the supercluster. It is perhaps seventy million light-years away in a direction roughly perpendicular to the plane of our galaxy. The photograph shows only one part of the cluster, which ranges across some twenty million light-years in all and is home to at least two hundred fifty large galaxies and perhaps a thousand or more lesser ones. Its population of many spiral galaxies mixed with a few ellipticals is characteristic of irregular clusters like Virgo.

In contrast to the Virgo Cluster, with its many spiral galaxies rather loosely dispersed, the Coma Cluster (left) is composed largely of ellipticals and the dust-free SO galaxies, relatively crowded at the cluster core. These are the characteristics of "spherical" clusters of galaxies. Over one thousand large galaxies and perhaps ten thousand dwarves are to be found here, brushing by one another at average leeways of less

138

than one million light-years—a density far greater than that of the Local Group.

The two giant galaxies visible in the photograph are a binary pair. Each in turn is orbited by a host of less massive galaxies. A binary structure of this sort is found in many clusters of galaxies. Evidence has been found that the Coma Cluster is part of an extensive, chainlike supercluster that may itself be binary; its companion, a cluster designated A1367, is two hundred fifty million light-years away.

137 Just under five hundred million light-years distant, the Coma Cluster of galaxies is home to an estimated thirteen hundred major galaxies, and is itself apparently subsumed within a super-cluster whose galactic population numbers over twenty-five hundred.

138 The Virgo Cluster of galaxies, seventy million light-years distant, is the nucleus of a supercluster that contains our Local Group and many other clusters of galaxies.

139

The Hercules and Perseus Clusters

Irregular in form, the Hercules Cluster lacks the central concentration of galaxies found in spherical clusters like Coma. But it has a noticeable structure, something like that of a meandering riverbed. Seven hundred million light-years away, it is one of four clusters belonging to the Hercules Supercluster, an association that stretches for a distance of fifty million light-years.

Many rich clusters of galaxies generate considerable amounts of energy in radio and X-ray wavelengths. Part of this energy comes of course from their member galaxies: Elliptical galaxies make up the most powerful X-ray and radio sources, and spherical clusters, where ellipticals abound, often show up prominently in X-ray or radio telescopes. But some rich clusters radiate X-ray energy from the regions between the galaxies as well. The source is thought to be clouds of hot gas ejected by elliptical galaxies or stripped from spirals that passed through the central regions of their clusters; it is from the central regions that most of the X-ray radiation comes.

In the case of the Perseus Cluster (left), about one-fifth of the X-ray output is generated by a single galaxy, NGC1275. Much of the rest of the X-ray radiation is emitted from an area nearly as wide as the cluster itself, some three million light-years in diameter. Additional evidence that intergalactic clouds are responsible is found when radio telescopes are trained on the Perseus Cluster. They reveal long "wakes" trailing behind several of the galaxies in the cluster, as if the galaxies were churning through intergalactic gas like boats cutting through water.

139 The Perseus Cluster of galaxies, three hundred and fifty million light-years distant, appears to us through a thick foreground scrim of stars in our own galaxy.

140 In this photograph of the Hercules Cluster (right), reproduced in negative to make visible many of the fainter galaxies, the sharply defined points and those with optically produced spikes are stars. Virtually all the other objects—all that are not obviously stars—are galaxies.

VI/Galaxies and the Universe

...It is no one dreame that
can please these all....
—BEN JONSON

A Journey toward the Edge of the Universe

Now comes an end to our journeying. Millions of years have passed on the planet of our birth, decades aboard. Clusters of galaxies pass abeam and are recorded in the logbooks as once we recorded the passing of stars and later of galaxies. The time has come to turn the ship around and decelerate until it can be brought to rest on a planet. We owe this much to the younger generations, who never saw the earth and have known only this life of ceaseless exploration. But for us elders it is the beginning of the end. Deceleration will take a long time, and we cannot hope to live to see the day when our crew will step out onto planetary soil, under planetary skies.

On the day when the deceleration order is to be given, we few survivors of the original crew gather in the observation dome for a last look at the cosmos while our ship is at its peak velocity. Few visit the observation dome any longer—to swim among the galaxies is unremarkable to those who have known no other surroundings—but to our old eyes the view remains awesome and a little frightening. The time-dilation effect having sped up the workings of the spiral galaxies that lie ahead of us by a factor of several million, they spangle with the light of millions of newly formed stars, and still more brilliant supernovae flash and crackle across them by the hundreds.

The captain rises with difficulty and proposes a toast. "To the unattainable goal," he calls, his glass raised toward the dome and to the galaxies in array. "To the edge of the universe."

"Hear, hear," we respond. How often we have talked about the edge of the universe, mapped it with our telescopes,

saluted it with this same toast. The phenomenon is as familiar to us as our names.

When we look across space we are also looking back in time. At distances of up to a few billion light-years we see galaxies as they were recently in cosmic history, looking much like those that lie nearby. At distances of five to ten billion light-years we are seeing younger galaxies whose light set out on its journey when the universe was about half its present age. At distances approaching fifteen billion light-years what we see are the brilliant beacons of galaxies being formed; they pour out huge quantities of energy, by comparison to which the births of stars in the contemporary cosmos seem but a bland reminiscence, like firecrackers set off to celebrate the anniversary of a revolution. At these distances we are seeing the denizens of a young cosmos, all light and noise.

The captain, projecting into cosmic time the tendency of the old to aggrandize the historical, sometimes speaks of events fifteen billion years ago as if he had been alive then, rather than having only witnessed them vicariously by telescope. "That was when galaxies were galaxies," he likes to say. "The juice squeezed out of ten thousand stars in a year. Stars blowing up with every tick of the clock. Energy aplenty—you could singe your hair just by stepping outdoors—and galaxies crowded so close together there was scarcely room to pass between them. A pilot had to keep on his toes in those days."

If we search with our telescopes for galaxies more distant than those at some fifteen billion light-years, we see nothing. At these distances we are looking back to a time before the primordial stuff of the universe had cooled sufficiently to congeal into stars and galaxies. That is what we mean be the edge of the universe—a temporal threshold marking the point in cosmic history before which darkness prevailed. It is an edge not of space but of time, and to visit it we would need not our spaceship, but a timeship able to travel into the past.

"To the unattainable goal." The first mate echoes the toast. "Faster than the galaxies."

This is a traditional riposte, one that refers to the expansion of the universe. The farther away a given galaxy we observe, the faster it is receding from us (or us from it, as you prefer) as it takes part in the universal expansion. In all directions we see galaxies on the threshold of the universe hurtling away from us, trains out of the past running on rails of the past, their lights the markers of the unattainable past.

The captain orders the ship brought about.

"Faster than the galaxies," the first mate repeats. "Fast as light." A mathematical witticism contained in special relativity prescribes that the fuel bill to accelerate any particle of matter to the speed of light would be infinite, would include the conversion of everything, itself included, into energy.

"Maybe the young folks are right to want to stop," the captain says. "They probably figure that otherwise we'd go on forever, that we'd burn up the whole universe in order to cross it. They figure we'd be firing up the boilers with tables and chairs when we'd left not one star shining in the sky."

"Don't worry, Captain," says the mate. "There will always be space travelers."

"Always have been," the captain replies. "We were space travelers before we ever left earth. See that galaxy over there?" He extends a gaunt finger. "When they look at the Milky Way, don't they see it speeding away at ten percent the speed of light? And that galaxy over there, don't they see it moving off at twenty percent the speed of light? And those millions of galaxies off near the edge, don't they see our galaxy moving almost as fast as light itself, just as we see them? Aren't we teetering on the edge of the universe, so far as they're concerned? Isn't it our part of the universe that's young and blinding bright, so says the old light that left here so long ago and only now is reaching them?

"We are all space travelers, gentlemen. We are. They are. All are."

The galaxies wheel across the sky as the ship is turned end-for-end.

"Let us show a light," says the captain. He produces a kerosene-burning ship's lantern, a treasured antique. He lights the wick, replaces the glass, and holds the brass lantern up to the windows of the dome. Its yellow flame mingles with the light of the galaxies.

"In a moment this flame will belong to our past," he says. "But it belongs to their future. Maybe one day an astronomer in one of those galaxies whose telescope is pointed the right way at the right time will catch this flicker from our little lantern. Just a couple of million miles' worth of light falling into the telescope, gone in a couple of seconds."

The captain blows out the lantern, sets it on the floor, takes the con, and gives the order to fire the engines.

"It's not so bad to be old, gentlemen," he says. "We're part of the future of most of the universe."

161

VI/GALAXIES AND THE UNIVERSE

Geometries of Space and Time

Albert Einstein once remarked, in a rare display of impatience over those who complain that the theory of relativity violates common sense, that "common sense" for each of us consists of what we learned prior to age sixteen. If we wish to improve our understanding of the cosmos we would do well to heed Einstein's remark, laying aside the prejudices of our common-sensical comprehension of things and taking to heart rules appropriate to the interstellar realm.

Prejudices dating from the childhood of our species run deep. For most of our history, we humans have tended to regard the earth as both stationary and central to the cosmos. This was an illusion. The earth is not stationary, but is moving in orbit about the sun, the sun in orbit about the center of the Milky Way Galaxy, the galaxy around the center of gravity of the Local Group, the Local Group in its orbit within the Local Supercluster, and the supercluster is moving as it participates in the expansion of the universe. Nor do we inhabit the center of the universe; no one does.

The cosmos is a study in motion and change. To set up a frame of reference in such conditions we are well advised to pay attention not only to *where* things are but *when*. The location of, let us say, the Rock of Gibraltar can be specified adequately using only the three dimensions of space, so long as our frame of reference is constrained to the surface of the earth: We simply say that it is located at the mouth of the Mediterranean, at about thirty-six degrees north by five degrees west. But these specifications no longer suffice once we step away from the earth and take a broader view. If we take neighboring stars as our reference point, the Rock of Gibraltar is being whisked along by both the rotational and orbital velocity of the earth; if we step back further and take an intergalactic perspective we must add in the sun's galactocentric velocity, and so forth.

In order to specify the location of Gibraltar in anything greater than a parochial context, we must therefore specify its location not only in the three dimensions of space but also in the fourth dimension of time. Relativity may be viewed as an attempt, and a highly successful one, to build this dictum into the foundations of physics. It does so by viewing events in a context of where-when called the space-continuum.

Four-dimensional geometry can present conceptual problems for creatures as visually oriented as we are, in that four-dimensional structures—a 4-D sphere, say, or a 4-D cube—are difficult if not impossible for us to visualize. A 4-D geometer must work without the aid of the mind's eye, like a perfumer imagining a fragrance yet to be created or Beethoven composing music from within the chambers of his deafness. These obstacles aside, viewing the universe in terms of a four-dimensional space-time continuum has made possible highly encouraging progress in cosmology, the science concerned with discerning the structure of the universe as a whole.

To see how the addition of a fourth dimension can deepen and refine our concepts of the cosmos, let us examine how adding a dimension can solve cosmological problems that arise on lower orders of dimensionality. Specifically, let us examine the dilemma of whether the universe is finite or infinite. How might this dilemma be resolved by creatures whose perceptions are limited to one-dimensional or two-dimensional frames of reference?

Imagine a one-dimensional world. Its citizens are known as Linelanders, a name that I have borrowed, like that of the Flatlanders to follow, from Edwin Abbott's lovely book *Flatland*. The actions and perceptions of Linelanders are confined to only two directions, forward and back. Each lives his life in a perpetual queue, just behind one Linelander and just in front of another. Passing muttered words up and down the queue, the Linelanders debate whether their cosmos is finite or infinite in extent. They know perfectly well that the *inhabited* world is limited; there is a Linelander who stands at the head of the line, and another who stands at its end. But beyond them the Line stretches into the distance. Does it ever end?

Some Linelander cosmologists maintain that it does not, that the Line is infinite in length. Their favorite form of argument is the reduction to absurdity. If the Line were finite, they argue, what would happen when one came to the end of it? How can there be an end to the cosmos? The notion is unthinkable; therefore the Line—the cosmos—must be infinite.

But Linelanders who favor the concept of a finite cosmos can point to equally troubling absurdities in the infinite cosmos model. One of their favorite lines of argument goes like this: If the Line is infinitely long, then where is its center? Why, nowhere. Or everywhere. Indeed, there is no universally agreeable way to designate points of reference on an infinite line. Yet here we are, standing on a given point, and there is another point on the Line a foot away; for this to be the case must not the Line be finite? One old Linelander cosmologist likes to refute the infinite-universe cosmologies by scoring the Line with a chalk mark and echoing Samuel Johnson's refutation of Berkeley: "I refute it thus!"

This is the dilemma of the finite versus the infinite universe. It was an ancient ponderable when Lucretius wrote about it in the first century B.C. And it appears to be a genuine dilemma, insoluble within the scope of dimensionality available to the intuitive apprehensions of the Linelanders.

Lineland

Finite but Unbounded Lineland

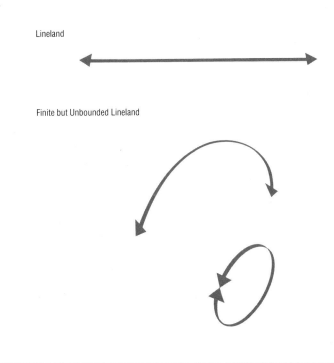

Figure 12. A Finite but Unbounded Lineland
The inhabitants of Lineland, limited in motion and perception to forward and backward, contemplate the dilemma of whether the cosmos is finite or infinite. An infinitely long Line seems unthinkable, but no less troubling is the question of how, if the Line is finite, it comes to an end. By introducing a second dimension—up and down—a Lineland cosmos may be created that though finite in extent has no boundary or edge. The finite-infinite dilemma is transcended.

But the finite-infinite dilemma for the Linelanders can be transcended if we add a dimension, permitting the Lineland cosmos to be bent.

Suppose we bend it into a loop. The result is that almost magic creation, a finite but unbounded universe. The Line has a finite length, yet it never comes to an end (Figure 12). A squad of Lineland explorers can be dispatched in any direction—remember, they have only two to choose from—and travel as far as they like without ever encountering an edge to their universe. If they travel far enough, they will find themselves back where they started. Their report reads: The universe is finite but unbounded; it has no edge and our choice of what to call its center is arbitrary. The debate is over and the cosmologists are free to turn their attention to deeper questions.

A similar transformation can be wrought upon the cosmos of two-dimensional creatures (Abbott's Flatlanders) by providing them with recourse to a three-dimensional cosmology. The Flatlanders live in a world of forward and back, right and left. So long as they confine their thinking to these two dimensions, they face a similar cosmological dilemma to that confronted by the Linelanders—either the plane is infinite in extent, or, equally unthinkable, it somewhere comes to an end. The dilemma can be evaded by introducing a higher dimension. Suppose we wrap the plane inhabited by the Flatlanders into the three-dimensional form of a sphere. Voila! The Flatlanders too have been bequeathed a finite but unbounded universe (Figure 13).

The Flatlanders, limited in their perception to two dimensions, cannot directly intuit the fact that they inhabit the surface of a sphere. But they can discover the shape of their spherical cosmos by means of experiment. If they dispatch an expedition of explorers to circumnavigate the globe, that feat will constitute convincing evidence that the Flatlander cosmos is spherical. A more subtle experiment can be performed by Flatland geometers without leaving home: They may lay out a triangle on the ground and measure the sum of its angles. Inflated by the swell of the globe, the angles will add up to more than the 180 degrees of a flat triangle. So Flatlander cosmologists, though unable to intuit or perhaps even to imagine a sphere, can deduce the spherical cosmos mathematically and intellectually.

We who think of ourselves as living in a three-dimensional world can accomplish similar feats by adding a fourth dimension to our conception of the cosmos. Working in four-dimensional geometries we can construct many models of our cosmos that are finite, in that they are composed of a

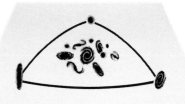

Figure 13. A Finite but Unbounded Flatland
For two-dimensional as for one-dimensional creatures, the introduction of a higher dimension overcomes the dilemma of a finite versus an infinite cosmos. Here a two-dimensional figure, a plane, is wrapped into the three-dimensional figure of a sphere, again resulting in a finite but unbounded universe.

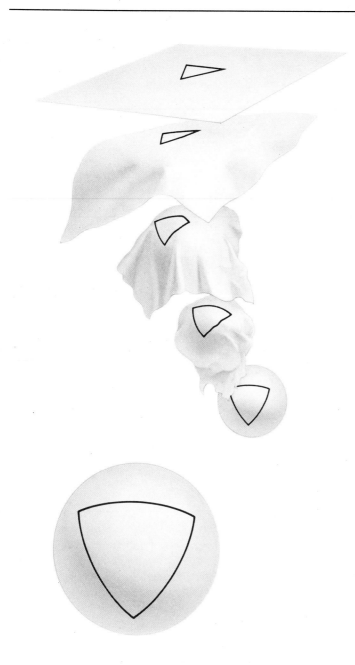

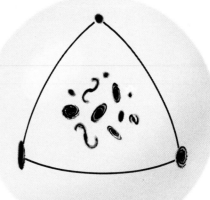

Figure 14. Spacetime Geometry Traced by Means of Light Beams
Lines that appear curved when viewed within the constraints of a given level of dimensionality may prove to be straight when viewed in terms of a geometry incorporating an additional dimension. In the two-dimensional universe represented schematically in the diagram at the top, a dense concentration of galaxies has affected the trajectory of light rays passing between galaxies A, B and C so that they appear curved to two-dimensional observers. But when a third dimension is added and the surface of the diagram is wrapped into a sphere (bottom), the same light rays are seen to be traveling by the shortest distance between A, B and C. In relativity, light rays that seem curved as observed by us three-dimensional observers actually are following the shortest path available along the contour of the four-dimensional space-time continuum.

finite number of galaxies and a finite amount of space separating them, and unbounded, in that one can travel in any direction indefinitely without reaching an edge. Many four-dimensional shapes of the cosmos may be imagined, but for the sake of simplicity let us stick to a spherical model and examine what might be some of the characteristics of a four-dimensional spherical universe.

To trace the geometry of a four-dimensional universe, let us use beams of starlight. This approach seems sensible since we can hope for no straighter line than a light beam, and it has the historical virtue of having been a favorite method of surveyors since the ancient Egyptians helped to invent geometry by employing sight lines to survey the boundaries of farms on flood plains of the Nile. Like Flatlanders tracing a triangle on the surface of their spherical cosmos, we can examine the four-dimensional structure of our universe by examining beams of starlight across large distances.

When we do this, we find that the light beams are not perfectly straight. They bend, and the degree to which they bend is directly related to their proximity to matter. A beam of light passing near a star curves toward the star and departs on an altered trajectory. Such a curvature of the space-time continuum was predicted by the general theory of relativity, and as it happened it was an observation of the bending of beams of starlight near the sun that constituted the first experimental proof of the theory.

Having established that light beams curve in the universe, we might say that space itself is curved. This is not an unreasonable statement, but to insist upon it is to needlessly assert the parochialism of our three-demensional prejudices. To see why, we rejoin the Flatlanders.

The Flatlanders who traced out a triangle on the surface of their spherical world might call the sides of that triangle curved, but we who can perceive the nature of a sphere see that they in fact represent geodesics, or the shortest distance between the points that they connect. International air travelers are familiar with this effect. A plot of the shortest air route for a round trip, say, from Los Angeles to Tokyo to Auckland and back to Los Angeles, perfectly efficient if plotted on a globe, will look curved if plotted on a flat map.

So it is theoretically possible to plot the shape of the cosmos in four dimensions by mapping the course of light beams passing across distances that make up an appreciable segment of the dimensions of the cosmos as a whole. Such distances are vast, and direct measurements of universal geometry in this fashion lie as yet beyond our reach, so that it

141

remains to be learned to which of the possible four-dimensional forms the contours of the universe most nearly conform.

Fortunately, we, like the Flatlanders, are not restricted to a single avenue of cosmological experiment, and can investigate universal geometries by other, less direct means. Since the fourth dimension of our four-dimensional paradigm is time, we may hope to learn something of the geometry of the universe by inquiring into its behavior over long periods of time.

---

141 This cluster of galaxies, barely visible among the bright foreground stars, is about five billion light-years distant. It represents the outer limits at which galaxies can be optically observed with existing telescopes and photographic equipment.

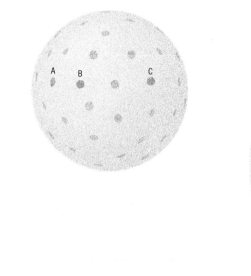

Figure 15. Expansion of the Universe
An expanding universe is here portrayed in terms of Flatlander cities arrayed upon the surface of a sphere. At each stage from left to right the size of the sphere has doubled, and so therefore have the distances between any two given cities. Since cities that were farther apart to start with must have traveled at higher relative velocities to double their distance in a given period of time, expansion reads out to any observer as a state in which the more distant a given city, the faster it is seen to be receding. By studying this phenomenon, Flatlanders can infer that they live in an expanding universe, as we infer expansion from the universal recession of galaxies. Note that for the Flatlanders the universe is not expanding *into* anything; it is simply stretching. The same may be said of our universe, which may be conceived of as a finite but unbounded system within which the density of the matter comprising the universe is decreasing as time goes by.

The Expansion of the Universe

Once Einstein had created, in relativity, a means of interpreting the cosmos in terms of four-dimensional geometry, he learned that his theory strongly implied that the universe was either expanding or contracting. Relativity would not, it seemed, permit the universe to remain static, but required that it either be opening like a blossoming flower or closing like a wilting one. This implication startled observers and theorists alike, Einstein not least among them. It had long been understood that there was plenty of motion in the cosmos, but the idea that the cosmos as a whole was engaged in *coherent* motion was radical. At first few took it seriously. Then astronomers discovered that remote galaxies are rushing apart from one another, and us from them, at velocities directly proportional to their distance—in short, that the universe is expanding.

To examine what is meant by an expanding universe, let us make a last visit to the world of Flatlanders on a sphere (Figure 15). Dotted across the surface of the sphere are Flatlander cities; these may be taken to represent clusters of galaxies in the universe. Now imagine that the sphere is being inflated.

Three intriguing characteristics of an expanding universe can be demonstrated using this model.

First, Flatlanders residing in any given city find that every other city is rushing away from them. This is true for every Flatlander, no matter in which city he lives. Each may choose to regard his city as being at rest, or another city as being at rest and his in motion, or all in motion; the choice is arbitrary.

Second, the rate at which the cities are rushing apart is directly proportional to their distances. Consider three cities, A, B and C. At the onset of our period of observation, A and B are one hundred miles apart, while B and C are two hundred miles apart. One unit of time later, the sphere has doubled in size. All intercity distances have doubled as well. A and B are now two hundred miles apart, B and C four hundred miles apart. We can see readily that for B and C to have opened up twice as much intervening space between

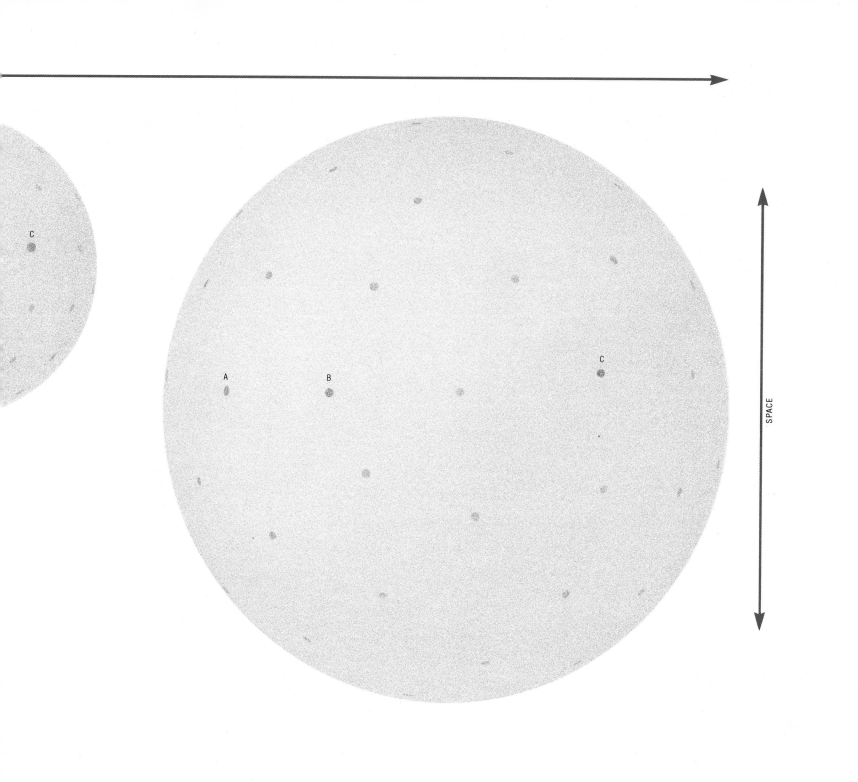

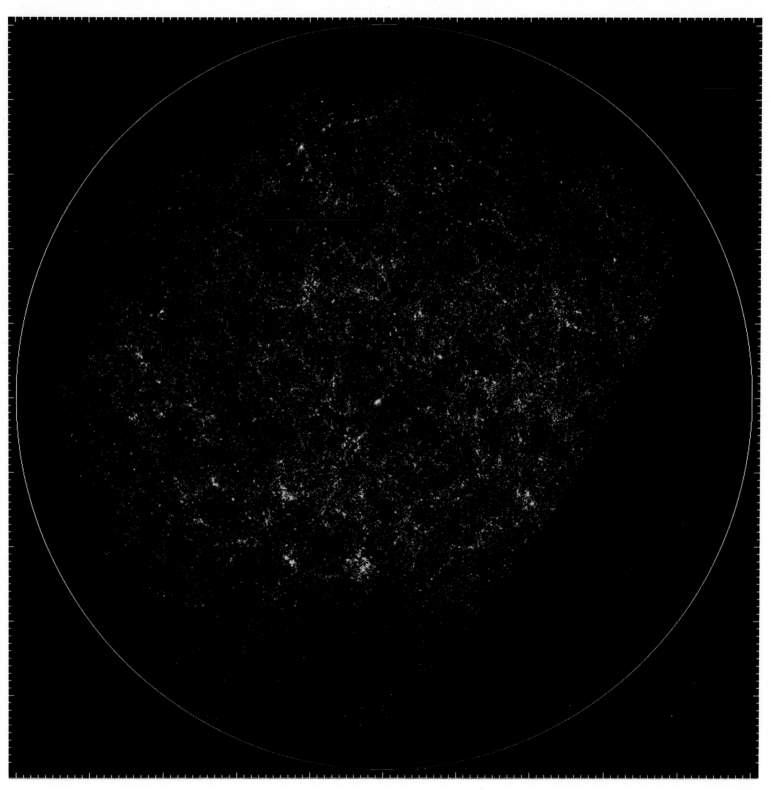

them as did A and B in the same period of time, the velocity of B relative to C must have been twice that of A relative to B. And the relative velocities of A and C, the two cities in our sample most distant from each other, must be still higher. That is the nature of expansion. Observers in any given Flatlander city can infer that their universe is expanding once they discover that all the cities are rushing apart at velocities directly proportional to their relative distances. It was by discovering this fact about clusters of galaxies that humans learned of the expansion of the universe.

A third characteristic of our expansion model is that no observer occupies a privileged or central position. No Flatlander city lies closer to the edge of the universe than any other, and none lies closer to the center of the universe, for on the surface of a sphere is to be found neither edge nor center. All observers are equally able to discover evidence of the expansion and to derive similar conclusions from it, and all observers see cities in all directions. Cosmologists call such a universe isotropic, meaning that its general appearance and behavior appear the same to all regardless of their location. This again seems to parallel the situation pertaining in the universe of galaxies: Every observer sees galaxies receding no matter in what quarter of the sky he looks.

It remains to be said that the shape of our universe is not necessarily, or even probably, analogous to a four-dimensional sphere. It may prove to be so, in which case it belongs to a class of "closed" geometrical forms, or it may adhere more closely to one of the "open" four-dimensional forms. If the universe is destined to stop expanding one day in the future, it is said to be closed, in other words to conform to a geometry analogous to that of a sphere. If instead the expansion of the universe is destined to go on forever, its geometry is said to be "open," or more nearly analogous to one of the various hyperbolic four-dimensional forms, usually described as saddle-shaped. Or the cosmos may prove to fit more closely to the template of a more exotic geometric figure, perhaps one involving still higher orders of dimensionality.

Since one of the dimensions of the space-time continuum is time, it is possible to theorize about the shape of the continuum by reasoning from information about how the expansion of the universe has proceeded during the course of cosmic history. The expansion of the universe, if engendered in a violent "big bang," ought not to have proceeded in a wholly unbridled fashion. The gravitational pull exerted by the clusters of galaxies upon one another ought to have retarded the rate of expansion to some degree. Cosmological models may be constructed in which the geometry of the cosmos is deduced from the deceleration rate.

How then do we investigate the fate of the universe? An ideal way would be to peer into the past. If we could observe how rapidly the universe was expanding billions of years ago, we could then compare that rate with measurements made locally of its rate of expansion today, and from the difference between the two determine the deceleration rate and predict whether or not expansion will go on forever.

Here, as in so many other ways, the universe is happily accommodating. Intergalactic space is so transparent that we can see galaxies billions of light-years away. And since light coming from those great distances has taken billions of years to reach us, we see the universe there as it was long ago. It is possible to study cosmic history directly. This effect the astronomers call "lookback time."

Lookback Time

What we see in the sky is the past. Light falling upon the earth tonight from the star Sirius, 8.7 light-years away, is 8.7 years old. Light from the red star Antares, 520 light-years away,

---

142 How many galaxies are there? No human yet knows. A rough estimate, based upon counting galaxies in our galactic neighborhood and extrapolating for the universe as a whole, is one hundred billion. This map plots the location of one million galaxies, or roughly one in ten thousand of all those in the universe. Its lacy, membranous pattern represents the largest-scale structure yet glimpsed by the human eye.

---

dates from the fifteenth century. We see the Andromeda galaxy as it was in the first days of *Homo Erectus*, the galaxies of the Virgo Cluster as they were when coconut palms grew at the North Pole and terror cranes darkened the skies of Earth. Light from distant quasars set out on its journey to our telescopes before the earth had formed. To look across space is to look back in time. The history of the cosmos is arrayed in the sky for those who care to read it.

Some of the implications of this situation may be investigated by means of a diagram in which space is plotted as the vertical axis and time the horizontal (Figure 16). The "light

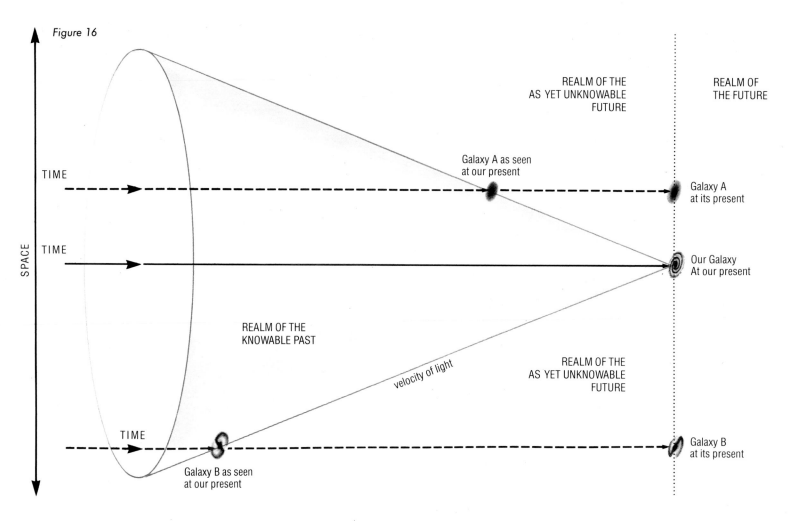

Figure 16

SPACE

TIME

TIME

TIME

REALM OF THE
AS YET UNKNOWABLE
FUTURE

REALM OF
THE FUTURE

Galaxy A as seen
at our present

Galaxy A
at its present

Our Galaxy
At our present

REALM OF THE
KNOWABLE PAST

velocity of light

REALM OF THE
AS YET UNKNOWABLE
FUTURE

Galaxy B
at its present

Galaxy B as seen
at our present

Figure 16. Intergalactic Past and Future Viewed in Terms of a Light Cone
Events occuring elsewhere in the cosmos make themselves known to us
only when their light (or other radiation) reaches us. Therefore, events
may be divided into those that lie within our "light cone," that is, events
whose light has had time to reach us, and those lying outside the light
cone, about which we can as yet have no knowledge. Our galaxy may
be envisioned as moving from left to right in this diagram as time
passes, so that events are constantly being swept into our light cone.

As the light cone diagram illustrates, the more distant the galaxy the
farther back in time we see it. Galaxy A, relatively nearby in space,
appears as it was in the recent past when its time line intersected the
boundaries of our light cone. Events that transpired in Galaxy A since
now belong to the history of Galaxy A, but lie in our future since they
have not yet entered our light cone. Galaxy B, more distant, intersects
our light cone at a point farther back in time and so is seen by us as it
was longer ago.

cone" in the illustration is created by drawing its sides along
a slope equal to the velocity of light, time against distance.
At any given point in cosmic history, the events that may be
known to a given observer are limited to those inside his light
cone (Figure 16).

If the universe were static and unchanging, the fact that we
can see back into its history would be of little use to us. But as
we live in a changing, evolving universe, lookback time

Figure 17. Light Cones of Observers in Three Galaxies
Here our galaxy and Galaxy A see each other in the relatively recent
past, while Galaxies A and B, farther apart, see each other in the more
remote past. Quite recent events in each galaxy are as yet unknown to
observers in other galaxies, the light announcing them not yet having
had time to traverse intergalactic space.

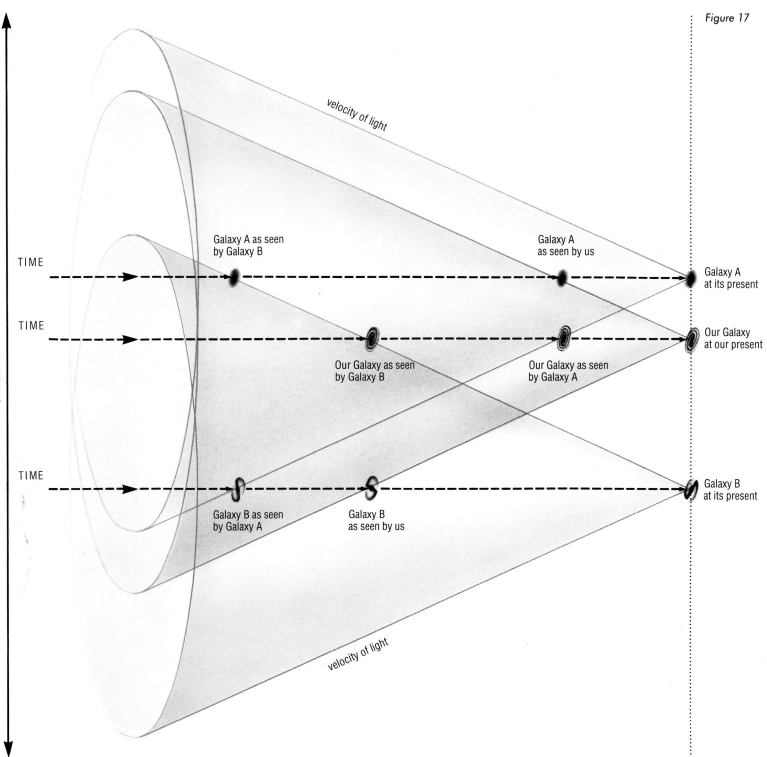

offers an enormous potential for learning. We are the beneficiaries of a deep reservoir of universal history from which we have as yet drunk but a few sips.

That we have not drunk more deeply is due not to limitations of the universe but merely to the limitations of existing telescopes. The telescopes built to date are insufficiently sensitive to photograph in any detail normal galaxies at distances of more than a billion light-years or so. Since galaxies are estimated to be some twelve to fifteen billion years old, a severe if temporary limit has been placed on our ability to observe them at large lookback times, and so the general history of the universe remains for the present largely unknown.

It is possible, however, to observe some objects at great distances—the quasars. Distant and brilliant, quasars probably are young galaxies caught in a process of forming or while still in their infancy, throwing off tremendous quantities of energy. They are so bright that they can readily be observed with existing telescopes even at distances of ten or fifteen billion light-years. Indeed, they are so brilliant that they ought to be detectable at even greater distances. But beyond about fifteen billion light-years, none has been found.

The absence of quasars at distances (and lookback times) of over about fifteen billion light-years can be explained handily in terms of "big bang" cosmology. If, as this theory maintains, the expansion of the universe began in a violent event roughly eighteen to twenty billion years ago, then we would expect that this moment of violent genesis was followed by an epoch of darkness during which the thinning, cooling, primordial matter of the universe coalesced. Then stars and the nuclei of protogalaxies began to shine, and light reappeared in the cosmos. If quasars are the protogalaxies predicted by the theory, we should expect to find a great many quasars at lookback times of approximately twelve to fifteen billion light-years, but none at greater lookback times, for beyond that point we are peering into the time prior to when protogalaxies had begun to shine. It is this quasar cutoff point that is sometimes referred to as the "edge" of the universe. All observers in the cosmos today find that the cutoff point occurs at extreme lookback times. No observer today is any closer to the "edge" than any other, for the "edge" belongs to the past. And no observer finds quasars abundant nearby, at short lookback times, as the quasars too belong to the past, and presumably since have settled down to become the nuclei of more or less normal galaxies.

And what of the light from the big bang itself? The theory predicts that it should be observable today as a ubiquitous background glow, its energy stretched out by the expansion of the universe until it has been reduced to a temperature of only three degrees above absolute zero. At this ebb, its peak energy will have been shifted well down into radio wavelengths. Just such a background glow was discovered by radio astronomers in 1965 and has been studied many times since. So it may well be that we are looking all the way back into the cauldron from which our present-day universe arose.

Figure 18. The Expanding, Evolving Universe Viewed in Terms of Lookback Time

The history of the universe as currently reconstructed is portrayed here as segments in a contemporary observer's light cone. The recent and local cosmos appears toward the right; remote times and distances lie to the left side of the diagram. The violent beginning of the expansion of the universe—the so-called big bang—appears at the extreme left, approximately eighteen or twenty billion years ago, its outpouring of energy still observable today as a dim background glow at radio wavelengths. A period of darkness follows while the universe expands and cools until matter is able to condense into stars and galaxies. The nuclei of young galaxies then blaze forth; very likely these are what we see today as quasars. In the billions of years that follow, the young galaxies settle into normal galaxies like those we know today; violent nuclei become less common and so fewer quasars are observed. Notice that a very large observational "twilight zone" intervenes for billions of light-years, representing lookback times where existing telescopes can perceive quasars but not the much dimmer normal galaxies. When telescopes capable of probing through this twilight zone are built, it ought to be possible to investigate by direct observation the whole history of the formation and evolution of galaxies.

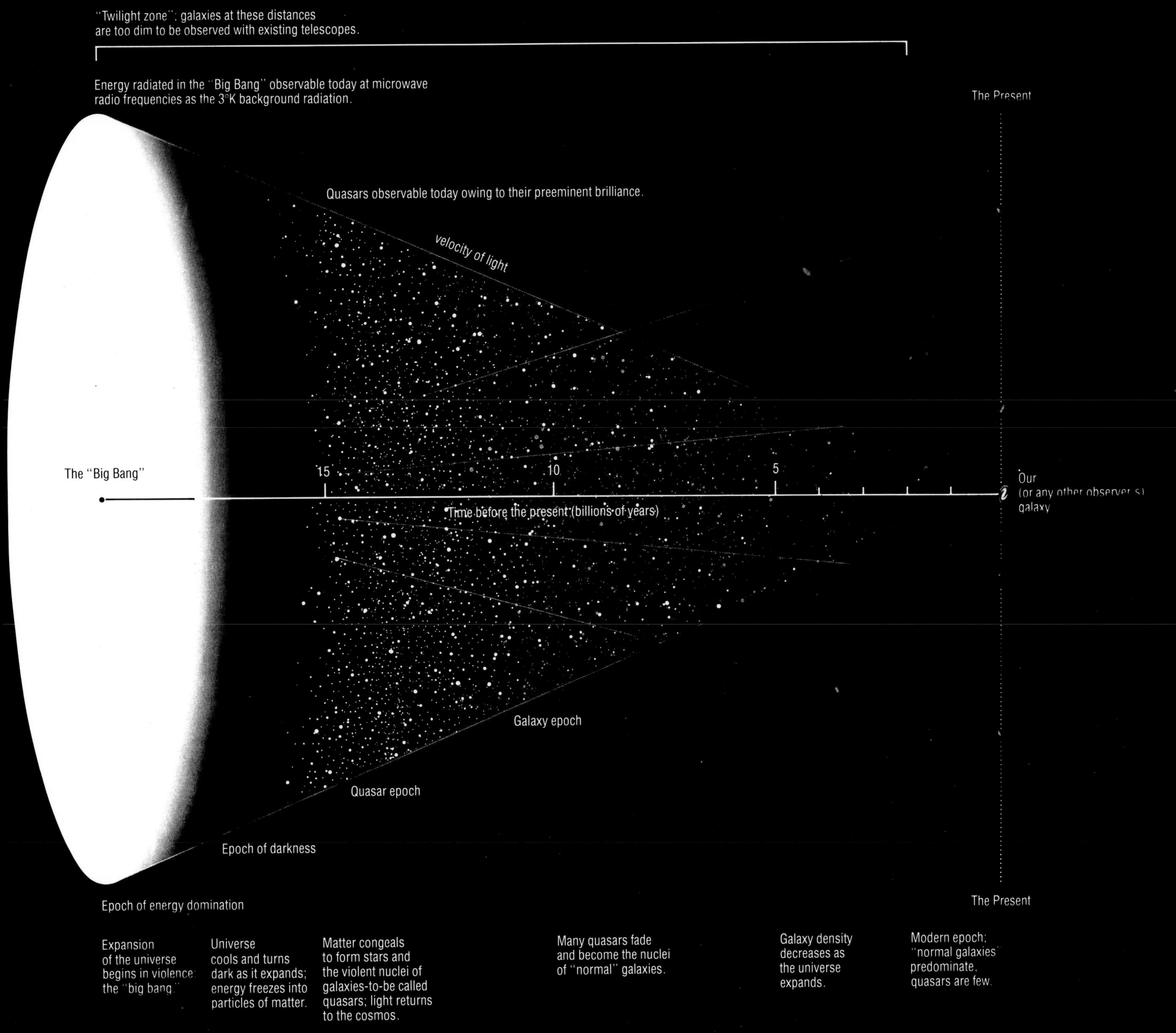

"Twilight zone": galaxies at these distances
are too dim to be observed with existing telescopes.

Energy radiated in the "Big Bang" observable today at microwave
radio frequencies as the 3°K background radiation.

The Present

Quasars observable today owing to their preeminent brilliance.

velocity of light

The "Big Bang"

15   10   5

Time before the present (billions of years)

Our
(or any other observer's)
galaxy

Galaxy epoch

Quasar epoch

Epoch of darkness

Epoch of energy domination

The Present

Expansion of the universe begins in violence: the "big bang."	Universe cools and turns dark as it expands; energy freezes into particles of matter.	Matter congeals to form stars and the violent nuclei of galaxies-to-be called quasars; light returns to the cosmos.	Many quasars fade and become the nuclei of "normal" galaxies.	Galaxy density decreases as the universe expands.	Modern epoch; "normal galaxies" predominate, quasars are few.

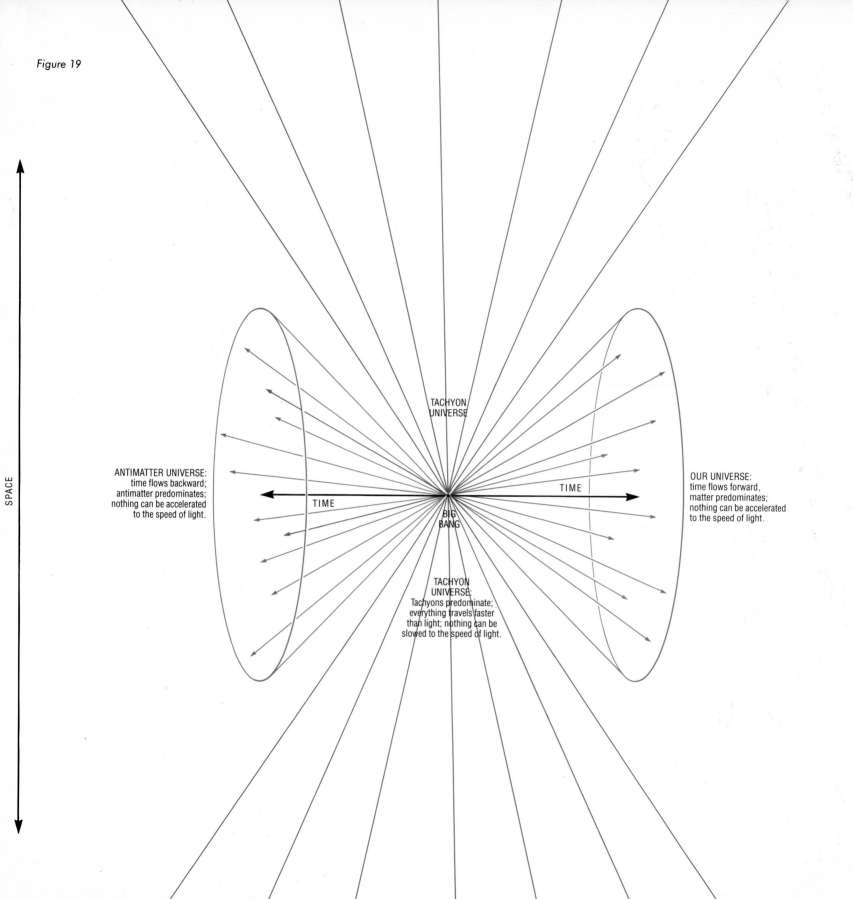

Figure 19

SPACE

TACHYON
UNIVERSE

ANTIMATTER UNIVERSE:
time flows backward;
antimatter predominates;
nothing can be accelerated
to the speed of light.

TIME

BIG
BANG

TIME

OUR UNIVERSE:
time flows forward,
matter predominates;
nothing can be accelerated
to the speed of light.

TACHYON
UNIVERSE:
Tachyons predominate;
everything travels faster
than light; nothing can be
slowed to the speed of light.

Modeling the Universe

A universe of questions about the universe remains. Evidence such as the quasar cutoff point and the cosmic background radiation indicate that those cosmologists are on the right track who maintain that the universe is finite in its age and material population, that its expansion began in an eruptive moment from which has unwound a process of cosmic evolution that continues today, and that the geometry of the universe may be more nearly understood by invoking orders of dimension beyond those that have served us locally here on earth. But within the boundaries of these assertions, even if they are quite correct, lies broad territory for cosmological investigation. Figure 19 illustrates just one of a multitude of imaginative cosmological models that may be constructed within the parameters of the big-bang hypothesis. The creation of J. Richard Gott III of Princeton University, this model proposes that not one but three universes sprang from the big bang. The Gott model attempts to account for two odd facts about our universe that trouble many cosmologists. One of these is that while the basic equations of physics are time-symmetrical—that is, they can be run forward or backward in time with equal efficacy—in the real universe time, it seems, moves in one direction only. The second oddity confronted by Gott is the scarcity of antimatter in our universe. For every sort of subatomic particle of matter, it is possible to conceive of a particle with the same mass but with opposite charge—an antiparticle. Yet only mere traces of antimatter have been found in nature. Why should nature be so nonsymmetrical, favoring matter over antimatter, running time in one direction but not the other?

Acting on a clue offered by theoretical physicists who maintain that antimatter can be thought of as ordinary matter moving in reverse time, Gott constructed his three-universe cosmology. He suggests that the big bang generated not only our universe, but also a second universe composed of antimatter and evolving in reverse time, as well as a third universe made up exclusively of particles that travel faster than light. The fleet particles of this ghostly third universe, called tachyons, are permissible under relativity theory, which requires only that nothing in our universe can be *accelerated* to the velocity of light; tachyons need not worry about this provision, for they have *always* been going faster than the speed of light. They occupy a mirror universe where everything travels faster than light and nothing can be reined to a velocity as slow as that of light. The Gott cosmology is a

masterpiece of symmetry without being dictatorial about it; it predicts, for instance, that there should be traces of contamination of our universe by antimatter (as has been verified by observation) and by tachyons (as has not). Whatever likelihood we might care to assign to its validity, in this combination of symmetry and imperfection the Gott model is redolent of nature's style.

The cosmological theories of today may be looked upon by our descendants with respect, bemusement, scorn or even hilarity, but the important thing is that cosmological endeavors need no longer be purely speculative in nature. We have learned how to test them against the real universe. And the universe is turning out to be remarkably amenable to such investigation. For all we knew until quite recently in human history, the cosmos at large might have been cramped or paltry, expressionless or opaque, unchanging or unpredictable. Instead, we find it lucid, intelligible, observable, evolving, and involving. The universe invites inquiry, as clear skies invite birds to fly.

Figure 19. A Three-Universe Big-Bang Cosmology
One of a number of sophisticated cosmological models that have been constructed within the broad purview of the big-bang account of cosmic history, this theory, proposed by J. Richard Gott III of Princeton University, postulates the existence of not one universe but three. It envisions that the big bang gave rise not only to our universe, where matter predominates over antimatter and time runs forward, but a second universe where antimatter predominates and time moves backward, as well as a tachyon universe where everything moves faster than the velocity of light. Our universe and the antimatter universe are segregated in terms of time. Both are segregated from the tachyon universe in terms of space, since the tachyons in the first instant of creation fled beyond the light cones of all observers in both the matter and antimatter universes.

143 (overleaf) Quasars are probably the nuclei of young galaxies caught in the act of shedding tremendous amounts of energy as they condense out of primordial gas. Here two quasars have been recorded at X-ray wavelengths by detectors aboard an earth satellite; the colors are not genuine, since X-rays lie well up the electromagnetic spectrum from light and so cannot be said to have color, but have been generated by computer. The dim quasar in the upper left corner is fifteen billion light-years away; its energy has been traveling through space for roughly three-quarters of the time since the expansion of the universe began.

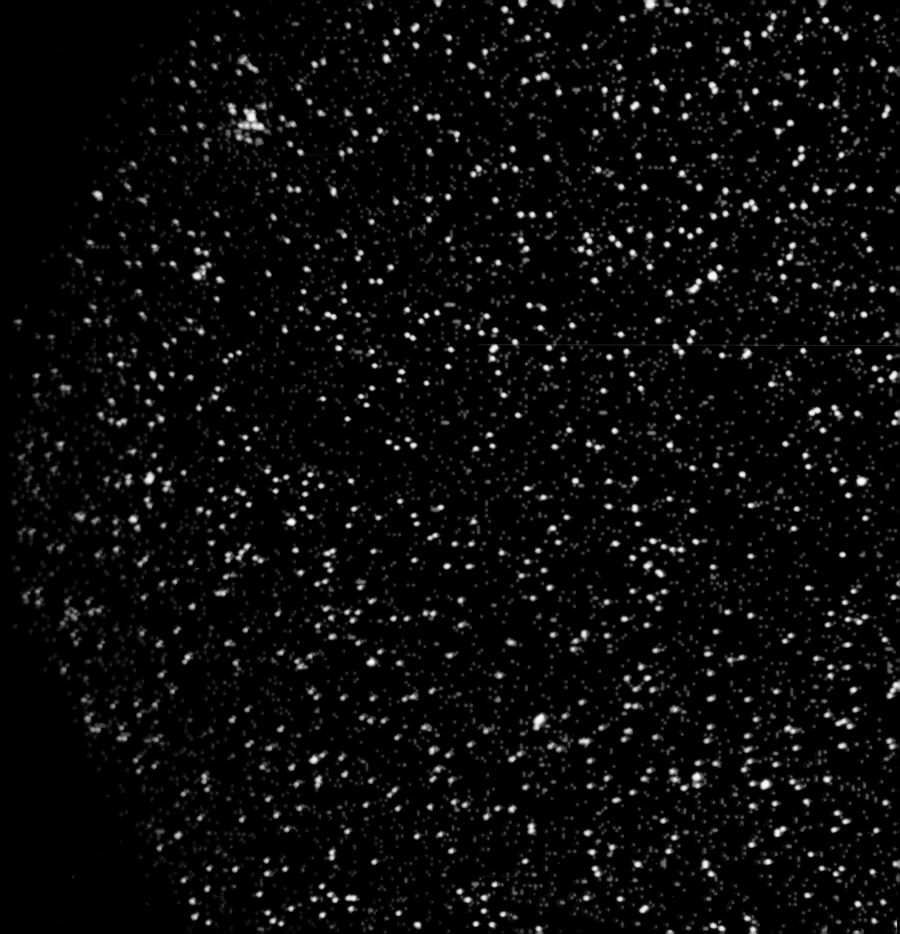

What we have learned
Is like a handful of earth;
What we have yet to learn
Is like the whole world.
—AVVAIYAR

Sources of Photographs

The author wishes to thank the following for their generosity in making their photographs available for this book. All photographs are copyright © the individuals or institutions named, and may not be reproduced without their permission.

Photo Number	Object	Source
Frontispiece	Galaxy NGC6744	Raymond J. Talbot, Jr., Reginald J. Dufour, and Eric B. Jensen, Rice University
1	The Sun	United States Naval Research Laboratory
2	Milky Way	Mssrs. Brodkorb, Rihm, and Rusche, Astrophoto Laboratory
3	Coalsack Nebula	Harvard College Observatory
4	Cone Nebula	Hale Observatories
5	Horsehead Nebula	Royal Observatory, Edinburgh
6	Orion Nebula	Royal Observatory, Edinburgh
7	Trapezium	Lick Observatory
8	Eagle Nebula	Hale Observatories
9	Rosette Nebula	Hale Observatories
10	Rosette Nebula, Interior	Kitt Peak National Observatory
11	Eta Carina Nebula	Association of Universities for Research in Astronomy, Inc., Cerro Tololo Inter-American Observatory
12	Eta Carina Nebula, Central Portion	Association of Universities for Research in Astronomy, Inc., Cerro Tololo Inter-American Observatory
13	Trifid and Lagoon Nebulae	Hale Observatories
14	Lagoon Nebula	Association of Universities for Research in Astronomy, Inc., Kitt Peak National Observatory
15	Trifid Nebula	David Malin, Anglo-Australian Telescope Board
16	Star Cluster NGC3293	David Malin, Anglo-Australian Telescope Board
17	Pleiades Star Cluster	Hale Observatories
18	Globular Star Cluster	Mt. Stromlo and Siding Spring Observatories, Australian National Observatory

Photo Number	Object	Source
19	Globular Star Cluster M13	United States Naval Observatory
20	Globular Star Cluster M3	Hale Observatories
21	Globular Star Cluster M15	Kitt Peak National Observatory
22	Globular Star Cluster Omega Centauri	Cerro Tololo Inter-American Observatory
23	Globular Star Cluster M5	Kitt Peak National Observatory
24	Globular Star Cluster 47 Tucana	Cerro Tololo Inter-American Observatory
25	Globular Star Clusters NGC6522 and NGC6528	Kitt Peak National Observatory
26	Intergalactic Globular Star Cluster NGC2419	Rene Racine, Hale Observatories
27	"Planetary" Nebula M27	Hale Observatories
28	"Planetary" Nebula M57	Hale Observatories
29	Veil Nebula Field View	Hale Observatories
30	Veil Nebula Detail	Kitt Peak National Observatory
31	Crab Nebula Superimposition	Guido Münch and Walter Baade, Hale Observatories
32	Crab Nebula	Lick Observatory
33	Milky Way in Cygnus	National Geographic-Palomar Observatory Sky Survey
34	Milky Way in Sagittarius	National Geographic-Palomar Observatory Sky Survey
35	Milky Way Mosaic	Mt. Stromlo and Siding Springs Observatories, Australian National University
36	Large Magellanic Cloud	Raymond J. Talbot, Jr., Reginald J. Dufour, and Eric B. Jensen, Rice University

Photo Number	Object	Source
37	The Magellanic Clouds	Harvard College Observatory
38	Small Magellanic Cloud	Royal Observatory, Edinburgh
39	Small Magellanic Cloud	Raymond J. Talbot, Jr., Reginald J. Dufour, and Eric B. Jensen, Rice University
40	Galaxy M31	Hale Observatories
41	M31 Central Regions	Association of Universities for Research in Astronomy, Inc., Kitt Peak National Observatory
42	Galaxy M31 Central Region	Hale Observatories
43	M31 Near Nucleus	Hale Observatories
44	M31 Outer Arms	Hale Observatories
45	M31 Nucleus	Lick Observatory
46	M32	Kitt Peak National Observatory
47	NGC205	Hale Observatories
48	NGC147	Hale Observatories
49	NGC185	Lick Observatory
50	M31 Radio Map	Elly M. Berkhuijsen, Max-Planck-Institut für Radioastronomie
51	M33	Hale Observatories
52	Sculptor Dwarf Spheroidal Galaxy	European Southern Observatory
53	M101	Hale Observatories
54	NGC7331	Hale Observatories
55-62	Galaxies by Formal Type	Hale Observatories
63	NGC2841	Hale Observatories
64	NGC2613	Allan Sandage, Hale Observatories
65	M64	Hale Observatories
66	NGC3623	United States Naval Observatory
67	M104	Steven Strom, Kitt Peak National Observatory
68	NGC4565	Lick Observatory
69	NGC3992	Lick Observatory
70	NGC4541	Steven Strom, Kitt Peak National Observatory
71	NGC1360	European Southern Observatory
72	NGC4650	Laird A. Thompson, Kitt Peak National Observatory
73	NGC4548	Laird A. Thompson, Kitt Peak National Observatory

Photo Number	Object	Source
74	M83	Raymond J. Talbot, Jr., Reginald J. Dufour, and Eric B. Jensen, Rice University
75	M83 Negative Print	Royal Observatory, Edinburgh
76	M83 Radio Map	D.H. Rogstad, California Institute of Technology
77	NGC4477	Steven Strom, Kitt Peak National Observatory
78	M84	Kitt Peak National Observatory
79	M49	Kitt Peak National Observatory
80	Carina Dwarf Galaxy	Royal Observatory, Edinburgh
81	NGC3077	Kitt Peak National Observatory
82	Sextans Dwarf Galaxy	Hale Observatories
83	NGC5364	Kitt Peak National Observatory
84	NGC6744	Cerro Tololo Inter-American Observatory
85	M74	Hale Observatories
86	NGC2683	United States Naval Observatory
87	NGC5907	United States Naval Observatory
88	Pluto & NGC5248	K. Alexander Brownlee
89,90	NGC4725 with Supernova	Hale Observatories
91	NGC4096 with Supernova	Lick Observatory
92	NGC4303 with Supernova	Lick Observatory
93	Centaurus A	Hale Observatories
94	Centaurus A	Raymond J. Talbot, Jr., Reginald J. Dufour, and Eric B. Jensen, Rice University
95	Perseus A	Hale Observatories
96	Cygnus A	Hale Observatories
97	M87 (large)	Malcolm Smith and W.E. Harris, Cerro Tololo Inter-American Observatory
98	M87	James Wray, McDonald Observatory, University of Texas
99	M87 (small)	Lick Observatory
100	M77	Lick Observatory
101	NGC1566	Harvard College Observatory
102	M106	James Wray, McDonald Observatory, University of Texas
103	M94	Kitt Peak National Observatory
104	NGC4151	Hale Observatories

Selected Bibliography

BOOKS ON GALAXIES AND RELATED SUBJECTS

Of General Interest

Abbott, Edwin, A., *Flatland*, New York, Dover Publications, 1952.

Allen, Richard Hinckley, *Star Names: Their Lore and Meaning*. New York, Dover Publications, 1963.

Berendzen, Richard Hart, and Daniel Seely, *Man Discovers the Galaxies*. New York, Science History Publications, 1976.

Bok, Bart J., and Priscilla F. Bok, *The Milky Way*, Cambridge, Mass., Harvard University Press, 1974.

Bondi, Hermann, *Relativity and Common Sense*. Garden City, N.Y., Anchor Doubleday, 1964.

Ferris, Timothy, *The Red Limit: The Search for the Edge of the Universe*. New York, William Morrow and Co., 1977.

Golden, Frederic, *Quasars, Pulsars, and Black Holes*. New York, Pocket Books, 1977.

Hoyle, Fred, *Galaxies, Nuclei, and Quasars*. New York, Harper and Row Publishers, 1965.

Maffei, Paolo, *Beyond the Moon*. Cambridge, Mass., MIT Press, 1978.

Mitton, Simon, ed., *The Cambridge Encyclopedia of Astronomy*. New York, Crown Publishers Inc., 1977.

Moore, Patrick, *The Amateur Astronomer*. New York, W.W. Norton and Company, 1968.

Page, Thornton, and Lou Williams Page, *Beyond the Milky Way: Galaxies, Quasars, and the New Cosmology*. New York, The Macmillan Company, 1969.

Page, Thornton, and Lou Williams Page, eds., *Stars and Clouds of the Milky Way: The Structure and Motion of Our Galaxy*. New York, The Macmillan Company, 1968.

Russell, Bertrand, *The ABC of Relativity*. New York, New America Library, 1970.

Scientific American, *New Frontiers in Astronomy*. San Francisco. W.H. Freeman and Company, 1975.

Shapley, Harlow, *Galaxies*. Cambridge, Mass., Harvard University Press, 1972.

Sullivan, Walter, *Black Holes: The Edge of Space, The End of Time*. Garden City, N.Y., Anchor Press/Doubleday, 1979.

Weinberg, Steven, *The First Three Minutes*. New York, Basic Books Inc. Publishers, 1977.

Whitney, Charles A., *The Discovery of Our Galaxy*. New York, Alfred A. Knopf, 1971.

Technical and Semi-Technical

Abetti, Giorgio, and M. Hack, *Nebulae and Galaxies*. New York, Thomas Y. Crowell Co., 1964.

Baade, Walter, *Evolution of Stars and Galaxies*. Cambridge, Mass., MIT Press, 1975.

Berkhuijsen, Elly M., and Richard Wielebinski, *Structure and Properties of Nearby Galaxies* (IAU Symposium no. 77). Boston, D. Reidel Publishing Company, 1978.

Clark, David H., and F. Richard Stephenson, eds. *The Historical Supernovae*. Oxford, Pergamon Press, 1977.

Dickens, R.J., and Joan E. Perry, eds., *The Galaxy and the Local Group* (Royal Greenwich Observatory Bulletin no. 182). Herstmonceux, Royal Greenwich Observatory, 1976.

Einstein, Albert, *The Meaning of Relativity*. Princeton, N.J., Princeton University Press, 1956.

Einstein, Albert, *Relativity: The Special and General Theory*. New York, Crown Publishers Inc., 1961.

Hazard, C., and S. Mitton, *Active Galactic Nuclei*. Cambridge, Cambridge University Press. 1979.

Hodge, Paul W., *The Physics and Astronomy of Galaxies and Cosmology*. New York, McGraw-Hill Book Company, 1966.

Lang, Kenneth R., and Owen Gingerich, eds., *A Sourcebook in Astronomy and Astrophysics, 1900-1975*. Cambridge, Mass., Harvard University Press, 1979.

Longair, M.S., and J. Einasto, *The Large Scale Structure of the Universe* (IAU Symposium no. 79). Boston, D. Reidel Publishing Company, 1978.

Middlehurst, Barbara M., and Lawrence H. Aller, *Nebulae and Interstellar Matter* (volume 7 of *Stars and Stellar Systems*). Chicago, University of Chicago Press, 1968.

Mitton, Simon, *Exploring the Galaxies*. New York, Charles Scribner's Sons, 1976.

North, J.D., *The Measure of the Universe: A History of Modern Cosmology*. Oxford, Oxford University Press, 1955.

O'Connell, D.J.K., ed., *Study Week on Nuclei of Galaxies*. Amsterdam, North-Holland Publishing Co., 1971.

Payne-Gaposchkin, Cecilia, *Stars and Clusters*. Cambridge, Mass., Harvard University Press. 1979.

Sandage, Allan, Mary Sandage, and Jerome Kristian, eds., *Galaxies and the Universe* (volume 9 of *Stars and Stellar Systems*). Chicago, University of Chicago Press, 1975.

Setti, Giancarlo, ed., *Structure and Evolution of Galaxies*. Boston, D. Reidel Publishing Company, 1975.

Shakescraft, John, ed., *The Formation and Dynamics of Galaxies* (IAU Symposium no. 58). Boston, D. Reidel Publishing Company, 1974.

Shapley, Harlow, *The Inner Metagalaxy*. New Haven, Yale University Press, 1957.

Shapley, Harlow, ed., *Sourcebook in Astronomy, 1900-1950*. Cambridge, Mass., Harvard University Press, 1960.

Shklovskii, Iosif S., *Stars: Their Birth, Life and Death*. San Francisco, W.H. Freeman & Co., 1978.

Shklovskii, Iosif S., *Supernovae*. New York, John Wiley and Sons, 1968.

Tayler, R.J., *Galaxies: Structure and Evolution*. New York, Crane, Russak and Company, 1978.

Tinsley, Beatrice M., and Richard B. Larson, eds., *The Evolution of Galaxies and Stellar Populations*. New Haven, Yale University Observatory, 1977.

Unsöld, Albrecht, *The New Cosmos*. New York, Springer-Verlag, 1977.

Woltjer, Lodewijk, ed., *Galaxies and the Universe*. New York, Columbia University Press, 1968.

PERIODICALS

Of General Interest

Astronomy. Milwaukee, Wisconsin, AstroMedia Corp.
Cosmic Search. Delaware, Ohio, Cosmic-Quest Inc.
Mercury. San Francisco, Astronomical Society of the Pacific.
Scientific American. New York, Scientific American Inc.
Sky & Telescope. Cambridge, Mass., Sky Publishing Corp.
Spaceflight. London, The British Interplanetary Society.

Technical

Annual Review of Astronomy and Astrophysics. Palo Alto, Calif., Annual Reviews Inc.
Astronomical Journal. New York, American Institute of Physics.
The Astrophysical Journal. Chicago, University of Chicago Press.
Journal of the Royal Astronomical Society. Oxford, Blackwell Scientific Publications.
Journal of the Royal Astronomical Society of Canada. Toronto, Royal Astronomical Society.
Monthly Notices of the Royal Astronomical Society. Oxford, Blackwell Scientific Publications.
Publications of the Astronomical Society of the Pacific. San Francisco, Astronomical Society of the Pacific.
Soviet Astronomy. New York, American Institute of Physics.
Vistas in Astronomy, New York, Pergamon Press.

TEXTBOOKS

Abell, George, *Exploration of the Universe,* 3rd ed. New York, Holt, Rinehart and Winston, 1975.
Field, George B., Gerrit L. Verschuur, and Cyril Ponnamperuma, *Cosmic Evolution: An Introduction to Astronomy.* Boston, Houghton Mifflin Co., 1978.
Hartmann, William K., *Astronomy: The Cosmic Journey.* Belmont, Calif., Wadsworth Publishing Company, 1978.
Motz, Lloyd, and Anneta Duveen, *Essentials of Astronomy,* 2d ed. New York, Columbia University Press, 1977.
Roy, A. E., and D. Clarke, *Astronomy: Structure of the Universe.* New York, Crane, Russak and Company, l977.

ATLASES AND CATALOGUES

Of General Interest

Becvar, Antonin, *Atlas of the Heavens.* Cambridge, Mass., Sky Publishing Corp., 1962
Becvar, Antonin, *Atlas of the Heavens–II. Catalogue.* Cambridge, Mass., Sky Publishing Corp., 1964.
Norton, Arthur P., and J. Gall Inglis, *Norton's Star Atlas and Reference Handbook.* Cambridge, Mass., Sky Publishing Corp., 1966.
Rey, H. A., *The Stars: A New Way to See Them,* 3d ed. Boston, Houghton Mifflin, 1967.

Sandage, Allan, *The Hubble Atlas of Galaxies.* Washington, D.C., Carnegie Institution of Washington, 1961.

Technical and Semi-Technical

Arp, Halton, *Atlas of Peculiar Galaxies.* Pasadena, California Institute of Technology, 1978.
Salentic, Jack W., and William G. Tifft, *The Revised New General Catalogue of Nonstellar Astronomical Objects.* Tucson, University of Arizona Press, 1973.
de Vaucouleurs, Gerard, and Antoinette de Vaucouleurs, *Reference Catalogue of Bright Galaxies.* Austin, University of Texas Press, 1964.
de Vaucouleurs, Gerard, Antoinette de Vaucouleurs, and Harold G. Corwin, Jr., *Second Reference Catalogue of Bright Galaxies.* Austin, University of Texas Press, 1976.
Zwicky, F., E. Herzog, and P. Wild, *Catalogue of Galaxies and Clusters of Galaxies.* Pasadena, Caltech University Press, 1960-1968.

EPIGRAMMATICAL MATERIAL

Blyth, R. H., *Zen and Zen Classics.* Tokyo, Hokuseido Press, 1964.
Horace, *Satires, Epistles, and Ars Poetica.* Cambridge, Mass., Harvard University Press, 1978.
Jonson, Ben, "The Vision of Delight," in W. H. Auden and Normal Holmes Pearson, eds., *Medieval and Renaissance Poets.* New York, Penguin Books, 1978.
Lao Tzu, *Tao Te Ching,* Gia-Fu Feng and Jane English, translators. New York, Alfred A. Knoft, 1974.
Lucretius, *De Rerum Natura,* Cyril Bailey, translator. Oxford, Clarendon Press, 1972.
Sagan, Carl, Frank Drake, Ann Druyan, Timothy Ferris, Jon Lomberg, and Linda Salzman Saga, *Murmurs of Earth: The Voyager Interstellar Record.* New York, Random House, 1978.
Shakespeare, William, "Romeo and Juliet," in *The Complete Works.* Baltimore, Penguin Books, 1969.
Waley, Arthur, *The Way and Its Power: A Study of the Tao Te Ching and Its Place in Chinese Thought.* London, George Allen and Unwin, 1968.
Whitney, Charles A., *The Discovery of Our Galaxy.* New York, Alfred A. Knopf, 1971.
Wilder, Thornton, *Our Town,* New York, Harper and Row, 1960.

PHOTOGRAPHS

Prints, slides, transparencies and in some cases posters of galaxies and other astronomical objects may be ordered from the following sources:

European Southern Observatory
c/o CERN
Geneva, Switzerland

Lick Observatory
University of California
Santa Cruz, California

Hale Observatories
Pasadena, California

United States Naval Observatory
Washington, D.C.

Kitt Peak National Observatory
Tucson, Arizona

Glossary

It's not easy to describe
the sea with the mouth
—Kokyū

Astronomical terminology can be confusing for those who are not trained in the field and far from perfectly lucid for those who are. Sometimes one word is applied to more than one sort of object; astronomers speak of interstellar "clouds" composed of gas and dust, but the Magellanic Clouds are galaxies and the term "cloud" may also mean a cluster of galaxies. Sometimes several names are applied to a single object, as when open star clusters are referred to as "galactic" clusters, or when the Local Group of galaxies is termed not a group but a cluster. Alternately, a single object may be bedecked with many designations. For example, the enormous galaxy M87 owes its name to the eighteenth-century catalogue of Charles Messier, but it is also known as NGC4486, after the New General Catalogue of 1888, 3C274 for the Third Cambridge Catalogue of Radio Sources, and as Virgo A, signifying that it is the most powerful source of radio energy in the constellation Virgo.

Misnomers abound. Many result from mistaken first impressions, like those that produced Columbus's "Indians." "Quasars" were so named because at first blush they look "quasi-stellar." By the time it became clear that they were probably not stars but rather the nuclei of distant galaxies the name had taken hold and it was too late to change it. The "planetary" nebulae are envelopes of gas disgorged by aging stars; first impressions to the contrary, they are about as unplanetary as anything can be.

I have tried to keep technical terminology to a minimum, and in one case I have knowingly suppressed important information in the interest of painting a simple picture: In the "journey" sections many of the perceptual distortions produced by relativistic space flight have been ignored, such as the blue-shifting of galaxies ahead and the red-shifting of those astern, in order to concentrate attention on the galaxies rather than on the effects of space flight. Technically-inclined readers are asked to forgive this omission.

Those impatient with technical terminology are invited to employ this glossary to cope with what little of it has proved unavoidable, and to take comfort in the reflection that the designations of galaxies, like the Latin names of plants, are but human inventions and remain unknown to that which they designate.

A. Denotes the most powerful source of radio energy within a given constellation, as in Cygnus A or Centaurus A.

Arp. Designates galaxies listed in the *Atlas of Peculiar Galaxies*, by the astronomer Halton Arp.

Atom. The smallest unit of matter of a chemical element. Atoms may be broken down into subatomic particles, but once this happens they will lose the chemical properties characteristic of their element. The repertoire of possible chemical interactions is vastly enhanced when atoms are combined into molecules. Many sorts of molecules, as well as free atoms, are found floating in space.

Barred Spiral. A spiral galaxy characterized by a prominent realm of stars and interstellar material arranged in the shape of a bar or spindle projecting out from either side of the central bulge. The bar is probably created by dynamical interactions in the overall gravitational environment of the galaxy. Most spiral galaxies show at least a trace of a bar. The term "barred" is reserved for those in which this feature is prominent.

Billion. The billion employed in this book is the American billion, equal to one thousand million, or 10^9.

Binary galaxies. A pair of galaxies bound together gravitationally. Normally they will coexist peacefully in orbit around their common center of gravity, but occasionally they will pass close to each other, producing spectacular distortions in their structures. The Milky Way and Andromeda galaxies form a binary pair.

Black hole. An object compressed to so high a density as to imprison even its own light. A black hole comes into existence when a collapsing star or other object wraps itself in a gravitational field sufficiently intense that the velocity required to escape from it exceeds the velocity of light, so that nothing can escape. Although the term invokes romantic imaginings of "holes in space," a black hole is at its heart quite a substantial object.

Bright nebula. See *Nebula*.

C. Designates radio sources listed in one of several Cambridge Catalogues of Radio Objects. Successive catalogues are designated by numbers preceding the letter designation, so that 3C273 means that the object —in this case a quasar—is number 273 in the Third Cambridge Catalogue.

Central bulge. The elliptical region at the center of a spiral galaxy, situated something like the yolk of a fried egg. Also called the lens.

Cepheid variable star. A supergiant pulsating star that varies in brightness. There are several classes of Cepheids, each valuable to astronomers in that the amount of time it takes them to go through a cycle of brightness variation is directly related to their intrinsic brightness. An astronomer seeking to determine the distance of a nearby galaxy can measure the period of variability of Cepheids there, derive their true brightness, compare this with their apparent brightness in the sky —the rule is that the apparent brightness of a star diminishes with the square of its distance—and so determine the distance of the star and of its galaxy. Cepheids are bright enough to be identified with existing telescopes in galaxies at distances of up to about ten million light-years. Polaris, the North Star, is a Cepheid variable.

Cloud. Alternate term for a cluster of galaxies. Also used informally for interstellar material within a galaxy, as in the "Monoceros Dark Cloud." The Magellanic Clouds are galaxies.

Cluster. An association of many galaxies bound together gravitationally. The gravitational bonds of a cluster of galaxies are strong enough to sustain the association against the expansion of the universe, so that the expansion takes place not within the cluster but in the spaces between clusters.

Constellation. A configuration of stars in the sky usually recognized as tracing out a recognizable figure or symbol. For convenience, modern star charts divide the entire sky into constellations. But constellations have little astrophysical significance since they tell us only where stars appear in the sky, not where they are relative to one another in real space. For example, the stars of Orion's belt are about sixteen hundred light-years away, while the distance to the star Betelgeuse at Orion's right shoulder is only five hundred twenty light-years, and Rigel, his left foot, is nine hundred light-years away.

Corona. See *Halo*.

Cosmic rays. Charged subatomic particles—most are protons—streaking through space at velocities approaching that of light. Eruptions on the surface of the sun are known to produce cosmic rays, as are supernovae, but other as yet unidentified sources are thought to exist.

Cosmology. The study of the structure and history of the universe at large. It can be subdivided into theoretical cosmology, which looks at the mathematical and physical possibilities of how the universe might be constructed, and observational cosmology, which gathers astronomical data relevant to cosmological questions. In practice, contributions to cosmology come from astronomers, astrophysicists, mathematicians and theorists working in a variety of fields and with a variety of styles.

Cosmos. A term for the universe that emphasizes the belief in an orderly underlying structure to all of creation. It comes from the Greek *Kósmos*, meaning 'order.'

Dark nebula. See *Nebula*.

Degenerate star. A star that has used up most of its nuclear fuel and has collapsed to a state of high density.

Density wave. In spiral galaxies, a wave propagated through the interstellar material of the disk in a spiral pattern. The wave promotes the collapse of interstellar clouds into new stars; the stars in turn light up the surrounding interstellar medium, creating the visible phenomenon we call the spiral arms. The density wave is thought to be generated by resonances in the gravitational interaction of the stars of the galaxy as they move in their orbits.

Diffuse nebula. See *Nebula*.

Disk. The flattened component of a spiral galaxy, home to billions of stars and to large tracts of interstellar material. See *Galaxy, spiral*.

Doppler shift. Displacement in the apparent wavelength of light or other radiation coming from a body that is in relative motion toward or away from the observer. If the object is approaching, its light will be compressed and will appear shorter in wavelength than if it were at rest. If it is receding, the opposite effect occurs and the light is shifted toward the long-wavelength, or red, end of the spectrum. Red shifts in the light of distant galaxies provide evidence that the universe is expanding.

Dwarf galaxy. See *Galaxy, dwarf*.

Dwarf star. This deceptively diminutive term is employed by astrophysicists to apply to most normal stars like our sun. In general usage, the term is often encountered with a modifier, as in white dwarf or black dwarf, which are degenerate stars that have collapsed to a size comparable to that of the earth.

Electron. A negatively charged subatomic particle that when found in an atom orbits the nucleus.

Electromagnetic spectrum. See *Spectrum*.

Elliptical galaxy. See *Galaxy, elliptical*.

Equator, galactic. The plane of the disk of the Milky Way Galaxy. The earth's equator is inclined sixty-three degrees relative to the galactic equator.

Event horizon. The boundary around a black hole from within which no matter or information can escape.

Evolution, stellar. The development of a star from its origin as a protostar, or recently collapsed ball of gas, through its career until it runs out of hydrogen and helium fuel and ebbs into darkness. For a normal star like our sun, this process takes billions of years, most of it spent on what astrophysicists call the "main sequence" where the star maintains a stable balance between the gravitational and radiative forces within it. When its fuel is exhausted, a star with the mass of the sun leaves the main sequence and expands enormously to become a "red giant" star. Then it ventilates much of its atmosphere into space and what remains of it settles down into a "white dwarf" star. The term "evolution" has been criticized by some researchers who point out that stars are not subject to Darwinian selection, but it remains useful in discussing stellar processes at large, if not the development of individual stars.

Fission, nuclear. See *Nucleus, atomic*.

Flare star. Dim dwarf stars that produce sudden, irregular outbursts of energy. They are probably stars that have recently formed and are still subject to imbalances between the gravitational force that tends to collapse them and the radiative pressure that tends to sustain them against collapse.

Fusion, nuclear. See *Nucleus, atomic*.

Galactic nucleus. See *Nucleus, galactic*.

Galactic star cluster. See *Star cluster, galactic*.

Galaxy. A giant association of stars and interstellar gas and dust. In mass, galaxies range from roughly ten million to perhaps ten thousand billion times that of the sun.

Galaxy, dwarf. A small, dim galaxy. Difficult to define with exactitude, dwarf galaxies are smaller than major galaxies like the Milky Way but larger than globular star clusters.

Galaxy, elliptical. A galaxy whose stars are arranged in an elliptical volume of space. Unlike the flattened spirals, ellipticals have no disk, no spiral arms and relatively little interstellar material. In shape they range from nearly spherical to almost cigar-shape.

Galaxy, irregular. A disorderly looking galaxy that displays little of the symmetry of ellipticals or spirals. Most irregulars are dwarves. Often they are satellites of larger galaxies.

Galaxy, spiral. A galaxy possessing a flattened disk marked by a pattern of spiral arms. In addition to stars, the disk contains interstellar clouds of gas and dust. The spiral arms are luminous areas within the interstellar medium where the clouds have been compressed sufficiently to trigger the foundation of stars, whose light in turn traces out the pattern of the arms.

Gamma rays. The highest-energy form of electromagnetic radiation, extremely high in frequency, short in wavelength. See *Spectrum*.

Gaseous nebula. See *Nebula*.

Globular star cluster. See *Star cluster, globular*.

Globule. A dark ball of interstellar dust and gas, often found in the vicinity of star-forming nebulae. In many cases, globules appear to be collapsing on their way to forming new stars. They have been described as dust balls rolled up in the turbulence of a collapsing cloud.

Gravitation. The universal attraction of particles of matter for one another. See *Gravity* and *Relativity, theories of*.

Gravity. The universal attraction of matter for matter. Like light and other radiation, the force of gravity decreases by the square of the distance, so that if the distance separating two galaxies is doubled, their gravitational attraction to each other will be reduced to one-quarter its original value.

Group. A small cluster of galaxies.

Halo. A spheroidal zone surrounding a galaxy and inhabited by old stars, globular clusters and clouds of gas. Also called the corona.

Helium. After hydrogen, the second simplest and second most abundant element in the universe.

H II region. A bright cloud of predominantly hydrogen gas that glows by virtue of its atoms having absorbed energy from nearby stars and re-emitted it, as occurs in a neon light. Normally these are regions where recently formed stars are pouring energy into the surrounding cloud from which they were created. In this book, H II regions are referred to by the broader term "bright nebulae." See *Nebula*, also *Hydrogen*.

Hubble constant. A measure of the rate of expansion of the universe. Modern estimates cite the Hubble constant at 50 kilometers per second per megaparsec. This means that for every megaparsec (i.e. 3.26 million light-years) farther out one looks, one finds galaxies receding at an additional 50 kilometers (31 miles) per second.

Hubble's law. The rule that the light from distant galaxies is red-shifted to a degree proportionate to their distance from us. Discovered by Edwin Hubble in 1929, this was the first indication of the expansion of the universe. The Hubble law is not valid within clusters of galaxies, which are gravitationally bound together and proof against the expansion of the universe, but comes into play in the realm of "pure Hubble flow" between superclusters of galaxies.

Hydrogen. The simplest and least massive of atoms, normally consisting of one proton and one electron. Hydrogen is by far the most common element in the universe. When a cloud of hydrogen gas has been ionized—that is, when many of its atoms have gained or lost electrons, as they will when energized by the radiation of a nearby star—it is referred to in astronomy as an H II region, after the chemical symbol for ionized hydrogen. Bright nebulae like the Orion Nebula in the Milky Way Galaxy and the Tarantula Nebula in the Large Magellanic Cloud are H II regions.

IC. Designates objects listed in the Index Catalogue, a supplement to the *New General Catalogue*.

Infrared light. Electronic radiation lying just to the low-frequency side of visible light on the electromagnetic spectrum; heat. Young stars still wrapped in the clouds from which they formed can sometimes be observed at infrared wavelengths. See *Spectrum*.

Interacting galaxies. Two or more galaxies that have drifted close enough together that their gravitational interaction has manifested itself in an obvious fashion, such as by structural distortions in each system or the exchange or expulsion of stars from them.

Interstellar medium. Matter found in the spaces between stars. In a normal spiral galaxy like ours, the interstellar medium is composed primarily of hydrogen and helium gas, traces of more complicated atoms and molecules, and dust contributed by the explosions of dying stars.

Irregular galaxy. See *Galaxy, irregular*.

Island universe. A galaxy. Though wonderfully descriptive of both the dimensions and independent stature of galaxies, the term has fallen from use in favor of others more concise and less presupposing.

Latitude, galactic. Coordinates specified in terms of degrees above or below the plane of the Milky Way Galaxy.

Lens. See *Central bulge*.

Lenticular galaxy. See *S0 Galaxy*.

Light-year. The distance traveled by light in one year. The velocity of light is 186,000 miles per second in a vacuum, and a light-year equals some 5.8×10^{12}, or nearly six thousand billion, miles.

Lookback time. A term employed to call attention to the fact that we see remote astronomical objects as they were when their light left them long ago. A galaxy one hundred million light-years away appears to us as it was one hundred million years in the past, while a quasar at ten billion light-years lookback time is seen as it was ten billion years ago, when the universe was perhaps one half its present age.

Longitude, galactic. Coordinates specified in terms of degrees along the plane of the galaxy measured eastward from the galactic center in the constellation Sagittarius.

M. Designates objects listed in the catalogue of Charles Messier, originally published in 1781. A comet-hunter, Messier listed in his catalogue any fuzzy-looking object that might be mistaken for a comet. As a result, the catalogue contains a polyglot mixture of bright and dark nebulae, open and globular star clusters, planetary nebulae and galaxies.

Magnitude. The brightness of a star or other astronomical object denoted on a logarithmic scale. A difference of five magnitudes equals a difference of 10^2 times in luminosity, while a difference of one magnitude equals a discrepancy in luminosity of 2.5 times. Objects brighter than zero magnitude are designated by minus numbers. The apparent magnitude of Sirius, the brightest star in the skies of earth after the sun, is minus 1.6; the North Star, Polaris, has a magnitude of 2; the dimmest stars visible to the unaided eye are about sixth magnitude. Large telescopes can detect objects at twenty-fourth magnitude and even dimmer.

Magnitude, absolute. The magnitude of a star or other astronomical body as it would appear if viewed at a standard distance of ten parsecs or 32.6 light-years. See *Magnitude*.

Magnitude, apparent. The magnitude of a star as it appears in the sky. See *Magnitude*.

Mass. The total amount of matter in a body. In this book, the masses of galaxies often are expressed in terms of their population of stars, for instance by stating that a galaxy has approximately one hundred billion stars like the sun. This assumes that the sun is a star of typical mass, a not wildly unlikely assumption. But if galaxies typically have a great number of low-mass dwarf stars too dim as yet to have been observed, as some astronomers suspect, then a galaxy with a mass one hundred billion times that of the sun would actually have many more than one hundred billion stars.

Microwaves. See *Spectrum*.

Milky Way. Our view of the disk of our galaxy, a softly glowing band of starlight stretching across the sky. By extension, our galaxy as a whole.

Nebula. Originally, any patch of hazy, diffuse light in the sky. A number of quite different objects fall under the umbrella of the term. Some are clouds of gas which have been excited to glow by hot stars within them. These are referred to as "bright nebulae" in this book. Astrophysicists designate them H II regions, after the chemical symbol for ionized hydrogen. Other bright nebulae consist of gas ejected from dying stars as "planetary" nebulae or, in the case of more violent stellar explosions, as supernovae remnants. H II regions, supernovae remnants and "planetary" nebulae are lumped together by astronomers under the term "gaseous nebulae." Interstellar clouds that glow not by emitting their own light but by reflected starlight are called "reflection" or "diffuse" nebulae. Clouds that do not glow at all are called "dark nebulae." Adding to the burden upon this single word is the fact that galaxies, in the days before telescopes could be built that were capable of resolving them into stars, also were classified as nebulae; even today one encounters the anachronistic term "spiral nebulae" for a spiral galaxy.

Neutron star. A degenerate star that has collapsed to extremely high density. A neutron star with the mass of the sun would measure only about twelve miles in diameter. Pulsars are believed to be neutron stars emitting energy at radio and other wavelengths as they spin; the energy spirals outward through the magnetic field of the pulsar like the jet of a water sprinkler, showing up to an outside observer as a pulse each time the streamer flashes past.

NGC. Designates objects listed in the New General Catalogue of Nonstellar Astronomical Objects.

Nova. An explosion of a star powerful enough that it dramatically if briefly increases the brightness of the star but mild enough so that it leaves a working star behind afterward. Novae were described by the late Cecelia Payne-Gaposchkin of Harvard University as "probably very old stars that are taking a drastic way out from an intolerable state, when they can no longer support themselves in the style to which they have been accustomed."

Nucleus, atomic. The center of an atom, around which whirl clouds of electrons. Considerable amounts of energy are employed in binding together the particles of the nucleus. Fission reactors and the atomic bomb work by breaking down the nucleus and releasing some of this energy; hydrogen bombs and stars release still more energy by fusing nuclei together.

Nucleus, galactic. The center of a galaxy. Galactic nuclei are typically small and bright. Their nature is not yet well known. Hypotheses concerning the anatomy of galactic nuclei range from dense star clusters to black holes.

Open star cluster. See *Star cluster, open.*

Parallax. The angle described by one astronomical unit—the mean distance from the earth to the sun—as viewed from a nearby star. Astronomers measure interstellar distances by photographing neighboring stars from alternate sides of the earth's orbit and measuring the displacement that this change of perspective introduces in their apparent location against the background of more distant stars.

Parsec. A unit of distance equal to 3.26 light-years. An abbreviation for parallax-second, a parsec is equal to the distance at which one astronomical unit—the mean distance between the earth and sun—describes an angle of one second of arc. Astronomers normally, for convenience, employ the parsec rather than the light-year as a unit of distance, and some argue that it would be simpler to expunge the latter term from the astronomical vocabulary. But both quantities are based upon the earth's orbit, arbitrary by transstellar standards.

Peculiar galaxy. A galaxy that fails to fit into one of the structural categories of spiral, elliptical, SO or irregular.

Planet. A body orbiting a star and shining by its reflected light. Planets up to about fifty times the mass of Jupiter might exist; in objects more massive than that, nuclear processes would begin at the core and they would become stars. A lower limit to the mass that a body must have to qualify as a planet has not yet been established, since the question has not come up here in the solar system and it is not yet possible to observe planets of other stars. In the solar system, the term "minor planet" or "asteroid" is applied to thousands of lesser bodies orbiting the sun, all of them much smaller than even the smallest planet, Pluto. Comets are objects of still lower density, most of which orbit the sun at great distances.

"Planetary" nebula. A shell of gas ejected into space by a star that has consumed much of its hydrogen fuel and has lost its internal balance. See *Nebula.*

Poles, galactic. The axis of the Milky Way Galaxy defined by drawing an imaginary line through the galactic nucleus perpendicular to the plane.

Proton. A heavy subatomic particle with a positive charge, found in the nuclei of atoms.

Pulsar. See *Neutron star.*

Quasar. A blue pinpoint of light starlike in appearance (hence the name, for Quasi-Stellar Object) but with a large red shift indicating that it is far away in the expanding universe. Quasars probably are the nuclei of young galaxies going through a violent stage during or immediately following their formation. Support has been lent to this theory by the discovery of distant galaxies with bright nuclei that closely resemble quasars.

Radio. Relatively low-frequency, long-wavelength electromagnetic radiation. The universe abounds in natural radio energy, much of it produced by atoms floating in interstellar clouds and by electrons being accelerated through space by magnetic fields. See *Spectrum.*

Radio galaxy. A galaxy that emits energy unusually strong in radio wavelengths.

Radio waves. See *Radio* and *Spectrum.*

Rattail galaxy. A pair of galaxies whose interaction has released large numbers of their stars and interstellar material in the form of a pair of distended plumes or tails.

Red shift. Displacement of spectral lines in the light of stars or galaxies toward the red or low-frequency end of the spectrum. Red shifts in the spectra of galaxies have been explained as representing the velocity of galaxies moving apart in the expansion of the universe. See *Spectrum.*

Relativity, theories of. Two theories of physics created by Einstein and based in part upon the recognition that in the absence of any universally authoritative frame of reference, any observer's frame of reference must be accepted as equally valid to any other. The special theory concerns bodies in uniform motion relative to one another; it derives such consequences as the equivalence of mass and energy ($E = mc^2$) and the apparent alteration in mass, shape and the rate of passage of time of objects in motion relative to an observer. The general theory, Einstein's theory of gravitation, envisions events taking place within a four-dimensional space-time continuum; stars and planets pursue geodesics—the shortest line between two points —along the continuum. Among its many other virtues, this approach does away with the necessity of invoking a "force" of gravity.

Ring galaxy. A galaxy shaped something like a smoke ring. It appears to be a transitional period in the life of a normal spiral galaxy induced by gravitational imbalances created when it collides with a smaller galaxy.

Satellite galaxy. A small galaxy orbiting a large one. The Magellanic Clouds are the largest of the numerous satellites of our galaxy.

Seyfert Galaxy. A galaxy with an unusually bright nucleus radiating strongly in blue and ultraviolet light. About one percent of major galaxies fit this category.

SO galaxy. A galaxy shaped like a spiral galaxy but lacking spiral arms and interstellar gas and dust.

Space-time continuum. See *Relativity, theories of.*

Spectroscope. A device for breaking down light or other radiation into its component frequencies. See *Spectrum.*
10382//bed

Spectrum. Electromagnetic radiation arranged in order of its wavelength. Unlike mechanical waves, electromagnetic waves can propagate through empty space. Their wavelengths range from as much as twenty miles for long-wave radio down to 5.5 trillionths of an inch for some gamma rays. Forms of electromagnetic energy listed here in order from longer to shorter wavelengths are: radio waves, microwaves, infrared light, visible light, ultraviolet light, X-rays and gamma rays. Radio telescopes examine the electromagnetic radiation of the first two groups, optical telescopes the next three, and orbiting detectors are used for astronomical observation in X-ray and gamma ray wavelengths. In casual usage, "spectrum" most often refers to a breakdown of light; spectra are employed to analyze the composition and behavior of stars and other astronomical objects.

Spiral arm. The luminous spiral pattern in the disks of spiral galaxies that lends them their name. Spiral galaxies typically have two major arms, though these may be fragmented into exquisitely intricate patterns. See *Density wave* and *Galaxy, spiral.*

Spiral galaxy. See *Galaxy, spiral.*

Spiral nebula. A spiral galaxy. The term is an anachronism dating from the days before galaxies had been resolved into stars, when it was uncertain whether they were independent galaxies or whirlpools of gas within our own galaxy. See *Nebula.*

Star. A self-luminous body of gas sufficiently compressed for nuclear fusion to operate at its core.

Star cluster, galactic. See *Star cluster, open.*

Star cluster, globular. A spherically shaped association of stars, smaller than a galaxy. Many globular clusters are found in the halos surrounding galaxies.

Star cluster, open. An association of stars smaller, more loosely organized and younger than a globular cluster. Most open clusters are composed of stars that formed together and are destined to dissipate across space as the cluster slowly falls apart. Open clusters are found in the disks of spiral galaxies, where star formation takes place, and so are sometimes referred to as "galactic" clusters.

Supercluster. An association of clusters of galaxies. Superclusters do not appear to be gravitationally bound and so probably are being stretched out or torn apart as the expansion of the universe proceeds. See *Cluster.*

Supernova. The explosion of a star. Titanic in their force, supernovae range from ten thousand to a million times more powerful than novae. Most of the mass of the star is blasted into space, leaving behind only a dense, cinderlike core. Supernovae occur when a massive star runs out of fuel, can no longer retain the radiative pressure that has sustained it, and collapses, creating such extreme heat and pressure at the core that the star detonates like a giant thermonuclear bomb.

Supernova remnant. Material cast into space by the explosion of a star as a supernova. Often quite massive, these remnants may remain visible in optical and radio wavelengths for much longer than the few tens of thousands of years that a typical "planetary" nebula survives. See *Supernova*

Telescope. A device for gathering and focusing energy so that distant objects may be studied. Telescopes are designed in accordance with the wavelength of the radiation they are intended to collect. Large optical telescopes employ a glass mirror to bring light to focus. Radio telescopes gather the much longer waves of radio radiation with a metal dish or a mesh of wires.

Time-dilation effect. In relativity theory, the slowing of the passage of time on board a starship or other object moving close to the speed of light relative to the passage of time at the home port it left behind. Time-dilation reaches fifty percent at about ninety percent of the speed of light and increases dramatically at velocities greater than that. Enormous amounts of energy would be required to accelerate even a small ship to velocities approaching that of light.

Twenty-one centimeter radiation. Energy emitted spontaneously by free hydrogen atoms. The twenty-one centimeter wavelength lies in the radio band of the electromagnetic spectrum, at fourteen hundred twenty megacycles. Many sorts of atoms have been detected in space by virtue of the spontaneous energy emissions, but since hydrogen is the most abundant element in space, the twenty-one centimeter radiation of hydrogen is especially useful in astronomy. As the wavelength of the radiation is exact, the velocities of clouds of gas can be determined by measuring *Doppler shifts* in their radio emanations.

Ultraviolet light. Electromagnetic energy of higher frequency than visible light, lying just beyond the blue end of the visible spectrum. Extremely hot stars, such as those that have recently ejected their shells as "planetary" nebulae and are collapsing toward the white dwarf stage, are prominent sources of ultraviolet energy. See *Spectrum.*

Universe. Everything. Compare *Cosmos.*

Variable star. A star the brightness of which varies periodically. There are many sorts of variable stars, some quite useful to astronomers as distance indicators. See *Cepheid variable star.*

Violent galaxy. A galaxy producing unusually high emissions of energy. About one percent of major galaxies fit this category. Also sometimes called an exploding galaxy, a term that is misleading insofar as it suggests that the galaxy might be flying apart; at most, a violent galaxy ejects only a small portion of its mass into intergalactic space.

Visible light. See *Spectrum.*

X-rays. High-frequency, short-wavelength electromagnetic radiation. Known cosmic sources of X-ray radiation include hot clouds of intergalactic gas and putative black holes. See *Spectrum.*

Index

Italicized numbers indicate illustrations.

The text was set in Futura Book by
U.S. Lithograph, Inc., New York, New York.
The book was printed four-color offset by
Dai Nippon Printing Co., LTD., Tokyo, Japan
The book was bound by Dai Nippon Printing Co., LTD., Tokyo, Japan.